T0332419

Cosserat Plate Theory

This book presents the foundation and validation of the Cosserat Plate Theory, numerical experiments of deformation and vibration, and the unique properties of the Cosserat plates. Our approach incorporates the high accuracy assumptions of the Cosserat plate deformation consistent with the Cosserat Elasticity equilibrium equations, constitutive formulas, strain-displacement, and torsion-microrotation relations. The Cosserat Plate Theory is parametric, where the "splitting parameter" minimizes the Cosserat plate energy. The validation of the theory is based on the comparison with the three-dimensional Cosserat Elastostatics and Elastodynamics. The numerical results are obtained using the Finite Element Method (FEM) specifically developed to solve the parametric system of equations. The analysis of deformation of a variety of Cosserat plates shows the stress concentration reduction, higher stiffness of Cosserat plates, and the size effect related to the microstructure. The analysis of vibration of Cosserat plates predicts size-related properties of the plate vibration, the existence of the additional so-called Cosserat plate resonances, and the dynamic anisotropy, related to the dependency of the resonances on the microelement's shapes and orientations.

Lev Steinberg is a professor of mathematics at the University of Puerto Rico at Mayagüez, USA. He is the author of dozens of patents and research publications in the area of applied differential equations and mechanical engineering.

Roman Kvasov is a professor of Mathematics at the University of Puerto Rico at Aguadilla, USA. He is an expert in Scientific Computing and the Finite Element Method.

Cosserat Plate Theory

Lev Steinberg

Roman Kvasov

CRC Press
Taylor & Francis Group
Boca Raton London New York

CRC Press is an imprint of the
Taylor & Francis Group, an **informa** business

First edition published 2023
by CRC Press
6000 Broken Sound Parkway NW, Suite 300, Boca Raton, FL 33487-2742

and by CRC Press
4 Park Square, Milton Park, Abingdon, Oxon, OX14 4RN

CRC Press is an imprint of Taylor & Francis Group, LLC

Library of Congress Cataloging-in-Publication Data

Names: Steinberg, Lev, author. | Kvasov, Roman, author.
Title: Cosserat plate theory / Lev Steinberg, Roman Kvasov.
Description: First edition. | Boca Raton : CRC Press, 2022. | Summary: "The
  book presents the foundation and validation of the Cosserat Plate
  Theory, numerical computations, and the unique properties of the
  Cosserat plates"-- Provided by publisher.
Identifiers: LCCN 2022001310 (print) | LCCN 2022001311 (ebook) | ISBN
  9781032040233 (hbk) | ISBN 9781032040240 (pbk) | ISBN 9781003190264
  (ebk)
Subjects: LCSH: Elastic plates and shells. | Finite element method. |
  Micropolar elasticity.
Classification: LCC QA932 .S74 2022 (print) | LCC QA932 (ebook) | DDC
  624.1/776--dc23/eng/20220201
LC record available at https://lccn.loc.gov/2022001310
LC ebook record available at https://lccn.loc.gov/2022001311

ISBN: 978-1-032-04023-3 (hbk)
ISBN: 978-1-032-04024-0 (pbk)
ISBN: 978-1-003-19026-4 (ebk)

DOI: 10.1201/9781003190264

Typeset in Kp font
by KnowledgeWorks Global Ltd.

Publisher's note: This book has been prepared from camera-ready copy provided by the authors.

Dedicated to our mothers:
Meita and Liudmyla

Contents

Preface

Scientists, engineers, and graduate students are usually exposed to a variety of different plate theories based on the classical or Cosserat continuum mechanics. The focus of this book is the Cosserat Plate Theory developed by the authors at the University of Puerto Rico.

The book presents the foundation and validation of the Cosserat Plate Theory, numerical computations, and the unique properties of the Cosserat plates. The plate theory is based on the Variational Principles for the Cosserat Elastostatics and Elastodynamics. This approach incorporates the high accuracy assumptions of the Cosserat plate deformation consistent with the Cosserat Elasticity equilibrium equations, constitutive formulas, strain-displacement, and torsion-microrotation relations. The Cosserat Plate Theory is parametric, where the "splitting" parameter minimizes the Cosserat plate energy. The validation of the two-dimensional theory is based on the high level of agreement with the three-dimensional Cosserat Elastostatics and Elastodynamics.

The numerical results are obtained using the Finite Element Method (FEM) specifically developed to solve the parametric system of equations. The FEM simulations include the analysis of the clamped Cosserat plates of different shapes and perforations under a variety of loads. The numerical results are consistent with the size effect known from the Cosserat Elasticity. The numerical experiments also detect and classify the additional high resonance frequencies of the plate depending on the shape and orientation of the microelements. This book is unique in that it provides the validated Cosserat Plate Theory and describes the newfound distinctive static and dynamic properties of the Cosserat plates.

The book will be of interest to applied mathematicians, engineers, and scientists working in the academia, naval and airspace research industry, and national laboratories. In particular, those interested in thin-walled structures made of modern materials with microstructure will find it interesting to learn the influence of the properties of the Cosserat materials on the static and dynamic behaviors of the thin-walled structures. The book can be used as additional reading material in such advanced graduate courses as Continuum Mechanics, Solid Mechanics, and Plates and Shells.

Acknowledgments

The authors are grateful to Dr. Paul Castillo for the helpful discussions related to the development of the Finite Element Method. Dr. Lev Steinberg would like to thank the University of Puerto Rico at Mayagüez for granting him the sabbatical leave to work on the manuscript. Dr. Roman Kvasov would like to thank his wife, Arlenys Ramírez, for her patience, support, and the valuable suggestions during the preparation of the book.

Introduction

The Classical Theory of Elasticity is based on the idealized model of elastic continuum, where the body forces acting on the surface element are described by the force vector. This assumption leads to the introduction of the symmetric stress and strain tensors. The theory gives accurate description of the behavior of such construction materials as steel, aluminum, and concrete, when the strains lie in the elastic limits. However, some significant differences between the theory and the experiment can be observed when the gradient of the strain is large enough, as in the cases of the concentration of strain around holes. The microstructure of the material has a large impact on the experimental results, when for small wavelengths and high frequency oscillations. Finally, the Classical Elasticity does not give a fair description of the processes in granular media and when acoustic waves travel through crystals, polymers, and polycrystal structures.

Many modern materials possess certain microstructure (cellular solids, pores, macromolecules, fibers, grains, voids, etc.) and exhibit experimental behavior that cannot be adequately described by the Classical Elasticity. Cosserat Elasticity describes the influence of the microstructure on the deformation of the material and is a very useful framework for modeling solid composites (fiber, platelet, or particulate), porous media (including solids with fissures or microcracks), and suspensions (containing isometric or anisometric particles) [56], [39].

The Cosserat Theory of Elasticity assumes that the transmission of forces through the area element is carried out by means of force and moment vectors. This leads to the asymmetry of the stress tensor and the introduction of the couple Cosserat stress tensor. The introduction of three additional degrees of freedom for microrotations results in the asymmetric Cosserat strain and the introduction Cosserat torsion tensor. The symmetric part of the strain corresponds to the classic strain [8], [35].

The examples of the Cosserat solids include rocks, concrete, polymers, and different composites [39], [38]. The experimental observation of coupled rotational-translational modes in a noncohesive granular phononic crystal is reported in [4]. These elastic wave modes are predicted by the Cosserat theory and are not described by the Classical Elasticity [4].

Aluminum-epoxy composites, being widely used in aircraft and aerospace industries [32], [14], were found to be Cosserat materials. The values of the relevant parameters based on specimen of an aluminum-epoxy composite were investigated in [36], [37]. Grained composites and closed-cell polymenthacrylimide foams were investigated and shown to be Cosserat materials [37]. The Cosserat moments and rotations, in addition to forces and displacements, were included in the model of the behavior of reinforced concrete in [54]. The engineering properties of the lightweight aggregate were investigated in [52]. The human bones were reported to be Cosserat materials in [44], while the Cosserat Elasticity was successfully used for their stress analysis of [48].

In this book, we will make the numerical computations for the plates made of rigid polyurethane foam reported to be Cosserat material in [44]. Insulation materials made from polyurethane foam are used in the construction of large industrial buildings and therefore the prediction of the behavior of the polyurethane foam under bending stress is extremely important [1]. Polyurethane foam is also used in structural insulated panels widely used in walls, floor slabs, and roofs [31].

Elastic plates are flat solids bounded by two parallel planes (faces) and an orthogonal surface called boundary. The distance between the faces of the plate is called thickness. The dimensions of the faces are assumed to be much larger than the thickness of the plate: the width-to-thickness ratio of a plate is considered to be greater than 5. The mathematical descriptions of the deformation of the elastic plates are called plate theories. Plate theories reduce the problem of determining the displacements and stresses of the plate under load from three-dimensional problem to two-dimensional [51].

Nowadays, plates play a crucial part in a variety of branches of modern technology. Such widespread use of thin-walled structures arises from their fundamental properties: lightweight, high load-carrying capacity, and technological effectiveness [13]. For this reason plates are widely used in aeronautical and aerospace industry, where lightweight is essential (for example, aluminum alloy plates) [6]. Marine engineering makes use of elastic plates for the design of the hull in shipbuilding: floor and sealing plates, frames, side girders, keel and margin plates made of steel. Plates are being increasingly utilized in chemical engineering (plate type heat exchangers), steam systems (orifice plate steam traps), plant and process design, powerplants (waterwalls, boilers, super-heaters, steam pipes, columns), civil, structural and mechanical engineering, construction and industrial machinery, transporta-tion and mining equipment, heat exchangers, reaction vessels, evaporators, transfer piping systems (nickel plates), and many other practical applications [3], [7].

The interest in the theory of deformed surfaces and the first appearance of the mathematical description of plates can be traced back to the late eighteenth century. In 1776, Euler per-formed a free vibration analysis of certain plate problems. In his "Discoveries in the Theory of Sound" in 1787, Chladni described the experiments on various modes of free vibrations of plates [23], [10]. J. Bernoulli developed theoretical justification of Chladni ex-periments in 1789 and used the direct approach for the derivation of the governing differential equation, considering the plate being a two-dimensional deformable continuum [10], [17], [13], [18].

The beginning of the nineteenth century was marked by the development of the general plate equation based on the direct approach. In 1813, Lagrange added a missing warping term to the equilibrium equation in the work of Germain and thus pro-duced what is now known as Germain-Lagrange bending equa-tion [13]. In 1821, Navier proposed the Newtonian conception that let him to formulate the general theory of elasticity: elastic reaction arises from the variation in intermolecular forces, which result from shifts in the unchangeable structure of the molecular configuration [25].

Not long after the three-dimensional elasticity equations were developed, Cauchy (1828) and Poisson (1829) used them to for-mulate the plate bending problem. The new technique, essen-tially different from the direct approach used in the previous works, was based on the splitting of the unknown stress com-ponents and thus reducing the three-dimensional equations to a two-dimensional problem. Even though the obtained governing differential equation coincided with the Germain-Lagrange bend-ing equation, the Poisson approach and the introduced bound-ary conditions became a subject for much debate and were widely criticized [17], [13].

In 1850, the revolutionary contribution was made by Kirchhoff. In his thesis, he presented the basis of the classical bending theory of thin elastic plates and triggered the future widespread use of the theory in practice. The energy functional of the three-dimensional elasticity theory was successfully simplified, thus reducing the three-dimensional theory to a two-dimensional plate bending problem [13].

In order to derive the bi-harmonic Germain-Lagrange differential equation, Kirchhoff came up with the following assumptions: straight lines normal to the mid-surface remain straight after deformation, straight lines normal to the mid-surface remain normal to the mid-surface after deformation and the thickness of the plate does not change during deformation [19]. In 1888, based on these assumptions, Love developed what is now called the Kirchhoff-Love Plate Theory – two-dimensional model describing the deformation of thin elastic plates [15], [16]. Saint-Venant proposed an extension to the Kirchhoff's plate, which takes into account both stretching and bending. In 1899, Levy obtained the first solution of the Kirchhoff's differential equation for the case of rectangular plates [19].

The development of the aircraft industry, nuclear physics, and chemical industry invoked a lot of analytical research of plates. This led to many extensive studies in the area of plate bending theory, such as contributions to the theory of large deformations and the general theory of elastic stability of thin plates, theoretical and experimental investigations associated with the accuracy of Kirchhoff Plate Theory, simplification of the general equations for the large deflections of very thin plates, development of the final form of the differential equation of the large-deflection theory, investigation of the postbuckling behavior of plates, solution of plates subjected to nonsymmetrical distributed loads and edge moments, development of the general theory of anisotropic plates, and nonlinear plate analysis [13].

Since Kirchhoff Plate Theory assumes the normal to the middle plane remaining normal during deformation, it automatically neglects transverse shear strain effects. A rigorous system of equations, which takes into account the transverse shear deformation, has been developed only in the middle of the twentieth century by Mindlin and Reissner [11], [12], [46], [26].

One of the main advantages of Mindlin-Reissner Plate Theory is that it is able to determine the reactions along the edges of a simply supported rectangular plate, where classical theory leads to a concentrated reaction at the corners of the plate [19]. The study of the relationships between the classical theory and the Mindlin-Reissner theory has proved that the solution of the clamped Mindlin-Reissner plate approaches the solution of the Kirchhoff plate as the thickness approaches zero and that the maximum bending can reach up to 20% for moderate plate thickness [27], [22]. In addition, the numerical calculations of bending behavior of the plate of moderate thickness show high level of agreement between three-dimensional and Reissner models [5], [27].

The theory of Cosserat Elasticity takes into account the effect of microstructure on the deformation of the body and gives a more precise description than the Classical Elasticity. In the 1960s, Green and Naghdi specialized their general theory of Cosserat surface to obtain the linear Cosserat plate [2]. In 1967, Eringen was the first to propose a theory of plates in the framework of Cosserat Elasticity [8]. Eringen based his theory on the assumption of no variation of microrotations in the thickness direction and a technique that is very similar to the one used for Kirchhoff plate – integration of the three-dimensional equations [8], [27].

In 2007, Steinberg proposed to use the Mindlin-Reissner Plate Theory in the framework of the Cosserat Elasticity. The theory described the deformation of thin Cosserat elastic plates and took into account the transverse variation of microrotation. It was developed in the thesis of Madrid [33] and later published in [26]. The analytical solution for simple rectangular plate was obtained by Reyes in [49]. The numerical solution of bending of Cosserat elastic plates according to this theory is presented in [41].

In this book, we will focus on the series of Cosserat plate theories, starting with the static theory developed by the authors in 2013 [40]. The distinctive feature of the new theory is that the obtained solutions are parametric and depend on the "splitting parameter". This allowed the highest level of approximation to the original three-dimensional problem [27]. The theory provides the equilibrium equations and constitutive relations, and the optimal value of the minimization of the Cosserat plate elastic energy. One of the main contributions of [27] is the comparison of the proposed Cosserat Plate Theory and the three-dimensional Cosserat Elasticity. It was shown that the precision of the Cosserat Plate Theory is similar to the precision of the classical Mindlin-Reissner Plate Theory [11], [12].

The numerical modeling of simply supported plates is given in [42], where a rigorous formula for the optimal value of the splitting parameter was developed. As the Cosserat Elasticity parameters tend to zero the solution of the Cosserat plate was shown to converge to the Mindlin-Reissner plate. This implies that the Cosserat Plate Theory is a generalization of the classic Mindlin-Reissner Plate Theory.

The Finite Element analysis of the perforated Cosserat plates is given in [43]. The Cosserat Plate Theory shows the agreement with the size effect, confirming that the plates of smaller thickness are more rigid than it is expected from the Mindlin-Reissner Plate Theory. The extension of the static theory of Cosserat plates to the dynamic problems was published in [28]. The computations predict a new kind of natural Cosserat frequencies associated with the material microstructure. The dynamics experiments are compatible with the size effect principle reported in [42] for the Cosserat plate bending. An extension of the paper [28] for different shapes and orientations of micro-elements incorporated into the Cosserat plates was published in the chapter [29]. The numerical computations of the plate free vibrations showed the existence of some additional high frequencies of microvibrations depending on the orientation of microelements. The comparison with the three-dimensional Cosserat Elastodynamics shows a high agreement with the exact values of the eigenvalue frequencies.

Notations

x_i	Cartesian coordinates
B	Cosserat body
P	Cosserat plate
h	plate thickness
μ, λ	Lamé parameters
$\alpha, \beta, \gamma, \epsilon$	Cosserat elasticity parameters
ρ	material density
J_{ji} or $\mathbf{J}$	microinertia
σ_{ji} or $\boldsymbol{\sigma}$	stress tensor
μ_{ji} or $\boldsymbol{\mu}$	couple stress tensor
γ_{ji} or $\boldsymbol{\gamma}$	strain tensor
χ_{ji} or $\boldsymbol{\chi}$	torsion tensor
u_i or $\mathbf{u}$	displacement vector
ϕ_i or $\boldsymbol{\phi}$	microrotation vector
p_i or $\mathbf{p}$	linear momentum
q_i or $\mathbf{q}$	angular momentum
ε_{ijk}	Levi-Civita tensor
U_C	strain stored energy
U_K	stress energy
T_C	stored kinetic energy
T_W	work of inertia forces

$\mathcal{S}$	Cosserat plate stress set
$\mathcal{U}$	Cosserat plate displacement set
$\mathcal{E}$	Cosserat plate strain set
η	splitting parameter
p	pressure
ω	frequency
θ	angle of microelement orientation
$M_{\alpha\beta}$	bending and twisting moments
Q_α	shear forces
$Q_\alpha^*, \hat{Q}_\alpha$	transverse shear forces
$R_\alpha^*, \hat{R}_\alpha$	Cosserat bending and twisting moments
S_α	Cosserat couple moments
Ψ_α	rotations of the middle plane around x_α axis
W, W^*	vertical deflections of the middle plate
$\Omega_\alpha^*, \hat{\Omega}_\alpha$	microrotations in the middle plate around x_α axis
Ω_3	rate of change of the microrotation

CHAPTER 1

Cosserat Elasticity

1.1 SCOPE OF THE CHAPTER

This chapter introduces the Theory of Cosserat Elasticity. Special attention is given to the infinitesimal elastic deformations related to the Linear Cosserat Elasticity. The corresponding equilibrium equations, constitutive formulas, strain-displacement and torsion-microrotation relations are presented. The chapter also includes the discussion of the Principle of Minimum Potential Energy, the Variational Principle for Cosserat Elastostatics, and the Variational Principle for Elastodynamics.

1.2 COSSERAT ELASTICITY

The Classical Elasticity is based on the idealized model of elastic continuum, which considers the material bodies as the subsets of the Euclidean space. It describes the elastic body deformation by the displacements of the points of the body. The Classical Elasticity assumes that the internal body forces act on the internal surface elements of the body. This assumption leads to the introduction of the symmetric stress tensor. The significant differences between the classical theory and the experiments are observed for the materials with microstructure. Many modern materials possess certain inner structure and may contain cellular solids, pores, macromolecules, fibers, grains, voids, etc. Elastic properties of such materials cannot be adequately described by the Classical Elasticity [38], [56]. The Cosserat Elasticity is a theoretical attempt to address this deficiency of the classical theory.

DOI: 10.1201/9781003190264-1

The Cosserat approach incorporates the characteristics of the material microstructure. It considers the material body as a subset of the Cosserat space, where the microelements are defined as infinitesimally small neighborhoods of any body point. Unlike the Euclidean space, the Cosserat space considers the microelements together with their **shape** and **orientation**. Therefore, the Cosserat Continuum is the theoretical framework for the Cosserat Elasticity, where the displacements of the points and microrotations of the corresponding microelements define the Cosserat body deformation.

The Cosserat Elasticity also assumes that the internal forces are transmitted through the body surface elements, not only by the forces but also by the moments. This leads to the asymmetry of the stress tensor and to the complementary concept of the couple-stress tensor. The local dynamics characteristics of Cosserat bodies include densities of mass and microinertia of the microelements.

1.3 LINEAR THEORY OF COSSERAT ELASTICITY

In this book, we use the usual Einstein notation that implies summation over a set of indexed terms in a formula. When an index variable appears twice in a single term, it implies summation of that term over all the values of the index. The index that is summed over we will call a "dummy index"; the index that is not summed over we will call a "free index". We will assume that the expressions that contain Latin letters (i, j, k, etc.) as subindices are understood to be the components of the spatial vectors (i.e. vectors from $\mathbb{R}^3$) and thus take values in the set $\{1, 2, 3\}$. We will also assume that the expressions that contain Greek letters (α, β, γ, etc.) as subindices are understood to be the components of the plane vectors (i.e. vectors from $\mathbb{R}^2$) and thus take values in the set $\{1, 2\}$.

We use interchangeably the common notations for the gradient

$$\text{grad } \mathbf{f} = \nabla \mathbf{f} = \frac{\partial \mathbf{f}}{\partial \mathbf{x}} = \left\{ \frac{\partial \mathbf{f}_i}{\partial \mathbf{x}_j} \right\},$$

where $\mathbf{f}$ is a vector function of $\mathbf{x} = x_j \mathbf{e}_j$.

The following classical identity will be frequently used without mention:

$$\int_V \mathbf{T} \cdot \nabla \mathbf{u} dv = \int_{\partial V} \mathbf{Tn} \cdot \mathbf{u} da - \int_V \mathbf{u} \cdot \operatorname{div} \mathbf{T} dv$$

for a tensor $\mathbf{T}$ and a vector $\mathbf{u}$.

1.3.1 Cosserat Space

The Cosserat space $\mathbb{C}^3$ is an extension of the Euclidean space $\mathbb{R}^3$. We define the elements of the Cosserat space $\mathbb{C}^3$ as a set of ordered pairs of the form

$$(\mathbf{x}, \mathbf{R_x}(\varphi)).$$

Here $\mathbf{x}$ is an element of $\mathbb{R}^3$, $\mathbf{n}$ is a unit vector associated with the element $\mathbf{x}$, φ is the microelement orientation angle $\boldsymbol{\varphi} = \varphi \mathbf{n}$, and $\mathbf{R_x}(\varphi)$ is an orthogonal matrix defined as

$$\mathbf{R_x}(\varphi) = \mathbf{I} + (\sin \varphi)\mathbf{Z} + (1 - \cos \varphi)\mathbf{Z}^2,$$

where

$$\mathbf{Z} = -\varepsilon \cdot \mathbf{n} = \begin{bmatrix} 0 & -n_3 & n_2 \\ -n_3 & 0 & -n_1 \\ -n_2 & n_1 & 0 \end{bmatrix}.$$

and ε is the Levi-Civita tensor [21].

Let us define the reference microelement associated with $\mathbf{x}$ as a set of points $\{\mathbf{x} + \mathbf{dx}\}$ infinitesimally close to $\mathbf{x}$. Then, the actual Cosserat microelement, which represents the rotation of the reference element by the angle φ around $\mathbf{n}$, is a set of points $\{\mathbf{x} + \delta \mathbf{x}\}$, where

$$\delta \mathbf{x} = \mathbf{R_x}(\varphi)\mathbf{dx}.$$

We will use the notation $\{\mathbf{i}_i\} = \{\mathbf{i}_1, \mathbf{i}_2, \mathbf{i}_3\}$ for the orthonormal frame in $\mathbb{C}^3$.

1.3.2 Cosserat Body Deformation

We assume that a Cosserat body B is represented by an open subset the Cosserat space. Let the reference configuration of a Cosserat body B be chosen. Under the Cosserat deformation the Cosserat material point $\mathbf{x}$ with the orientation matrix $\mathbf{R_x}(\varphi)$ moves to a new point $\mathbf{x}'$ with a new orientation matrix $\mathbf{R_{x'}}(\varphi')$.

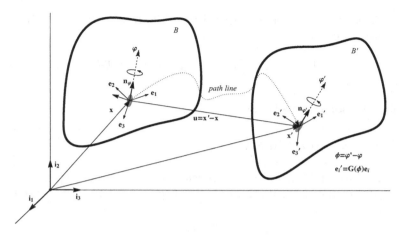

Figure 1.1   Deformation of a Cosserat body.

We describe the deformation by a sufficiently smooth vector function **f**, such that

$$\mathbf{x}' = \mathbf{f}(\mathbf{x}), \tag{1.1}$$

and by the matrix transform $\mathbf{G}(\phi)$

$$\mathbf{R}_{\mathbf{x}'}(\varphi') = \mathbf{G}(\phi)\mathbf{R}_{\mathbf{x}}(\varphi), \tag{1.2}$$

where ϕ is called the microrotation vector.

We also have

$$\mathbf{e}'_i = \mathbf{G}(\phi)\mathbf{e}_i, \tag{1.3}$$

where $\mathbf{e}_i$ and $\mathbf{e}'_i$ are the orthonormal frames associated with the material point $\mathbf{x}$ in the reference and current configurations (see Figure 1.1).

Let the deformation gradient **F** be defined as

$$\mathbf{F} = \nabla\mathbf{f}.$$

In the Classical Elasticity it is common to consider two infinitesimally close points $\mathbf{x}$ and $\mathbf{x} + \mathbf{dx}$ before deformation and the same points after the deformation $\mathbf{x}'$ and $\mathbf{x}' + \mathbf{dx}'$. Therefore, we have that

$$\mathbf{dx}' = \mathbf{F}\mathbf{dx}, \tag{1.4}$$

If we introduce the displacement vector **u** as

$$\mathbf{u} = \mathbf{x}' - \mathbf{x},$$

and the displacement gradient $\mathbf{F_u}$ as

$$\mathbf{F_u} = \mathbf{F} - \mathbf{I},$$

then (1.4) implies that

$$d\mathbf{u} = \mathbf{F_u}d\mathbf{x}. \tag{1.5}$$

We apply a similar approach to the Cosserat deformation. We pick two infinitesimally close points $\mathbf{x}$ and $\mathbf{x} + \delta\mathbf{x}$ before deformation and the same points after the deformation $\mathbf{x}'$ and $\mathbf{x}' + \delta\mathbf{x}'$, where

$$\delta\mathbf{x} = \mathbf{R_x}(\varphi)d\mathbf{x}, \tag{1.6}$$

$$\delta\mathbf{x}' = \mathbf{R_{x'}}(\varphi')d\mathbf{x}'. \tag{1.7}$$

From the equation (1.6) we have

$$d\mathbf{x} = \mathbf{R_x^{-1}}(\varphi)\delta\mathbf{x}, \tag{1.8}$$

Now we substitute the expression (1.8) into (1.5), and then into (1.7) to obtain

$$\delta\mathbf{x}' = \mathbf{R_{x'}}(\varphi')\mathbf{F}\mathbf{R_x^{-1}}(\varphi)\delta\mathbf{x}.$$

We can now introduce the Cosserat deformation gradient $\mathbf{\Phi}$ as follows:

$$\delta\mathbf{x}' = \mathbf{\Phi}\delta\mathbf{x},$$

where $\mathbf{\Phi}$ is the following product

$$\mathbf{\Phi} = \mathbf{R_{x'}}(\varphi')\mathbf{F}\mathbf{R_x^{-1}}(\varphi) = \mathbf{G}(\phi)\mathbf{R_x}(\varphi)\mathbf{F}\mathbf{R_x^{-1}}(\varphi).$$

Using the following definition of the Cosserat infinitesimal displacement

$$\delta\mathbf{u} = \delta\mathbf{x}' - \delta\mathbf{x},$$

we obtain that

$$\delta\mathbf{u} = \mathbf{\Phi_u}\delta\mathbf{x},$$

where

$$\mathbf{\Phi_u} = \mathbf{\Phi} - \mathbf{I}.$$

The tensor $\mathbf{\Phi_u}$ is called the Cosserat displacement gradient.

We define the Cosserat body motion similarly by the time dependent functions $\mathbf{f}(\mathbf{x}, t)$ and $\mathbf{G}(\phi, t)$.

1.3.3 Infinitesimal Cosserat Deformation and Motion

In the linear theory we assume that the orientation angles are infinitesimally small; therefore, the orientation matrices before and after the deformation can be written as

$$\mathbf{R}_x(\varphi) \approx \mathbf{I} + \varphi \mathbf{Z} = \mathbf{I} - \varepsilon \cdot \varphi,$$
$$\mathbf{R}_{x'}(\varphi') \approx \mathbf{I} + \varphi' \mathbf{Z} = \mathbf{I} - \varepsilon \cdot \varphi'.$$

The linear approximation of the matrix transform, therefore, is

$$\mathbf{G}(\phi) = \mathbf{R}_{x'}(\varphi')\mathbf{R}_x^{-1}(\varphi)$$
$$\approx (\mathbf{I} - \varepsilon \cdot \varphi)(\mathbf{I} + \varepsilon \cdot \varphi')$$
$$\approx \mathbf{I} - \varepsilon \cdot \phi,$$

where

$$\phi = \varphi' - \varphi, \tag{1.9}$$

is called the linearized microrotation vector.

Now we can find the linear approximation of the Cosserat deformation gradient

$$\boldsymbol{\Phi} = \mathbf{G}(\phi)\mathbf{R}_x(\varphi)\mathbf{F}\mathbf{R}_x^{-1}(\varphi)$$
$$\approx (\mathbf{I} - \varepsilon \cdot \phi)\mathbf{R}_x(\varphi)(\mathbf{F_u} + \mathbf{I})\mathbf{R}_x^{-1}(\varphi)$$
$$\approx (\mathbf{I} - \varepsilon \cdot \phi)(\mathbf{F_u} + \mathbf{I})$$
$$\approx \mathbf{F_u} + \mathbf{I} - \varepsilon \cdot \phi.$$

From the linear approximation of the Cosserat deformation gradient

$$\boldsymbol{\Phi} \approx \mathbf{F_u} + \mathbf{I} - \varepsilon \cdot \phi,$$

we can find the linear approximation for the Cosserat displacement gradient

$$\boldsymbol{\Phi_u} = \boldsymbol{\Phi} - \mathbf{I} \approx \mathbf{F_u} - \varepsilon \cdot \phi.$$

We will also use the gradient $\nabla \phi$ of the microrotation vector ϕ. It is convenient to use the transposed Cosserat deformation tensors

$$\gamma = (\boldsymbol{\Phi_u})^T,$$
$$\chi = (\nabla \phi)^T,$$

where γ is called the Cosserat strain tensor and χ is called the torsion tensor. These tensors can be written in the component form as the Cosserat strain-displacement and torsion-microrotation relations [55]:

$$\gamma_{ji} = u_{i,j} + \varepsilon_{ijk}\phi_k, \tag{1.10}$$

$$\chi_{ji} = \phi_{i,j}. \tag{1.11}$$

The symmetric part of the Cosserat strain tensor

$$\gamma_{\text{sym}} = \frac{1}{2}\left(\gamma + \gamma^T\right),$$

coincides with the strain tensor of the Classical Linear Elasticity.

1.3.4 Conservation of Mass and Microinertia

Let us consider a Cosserat body motion $B' = B(t)$. The conservation of mass and microinertia postulates that for any part $V' \subset B'$:

$$\frac{d}{dt} \int_{V'} \rho' dv' = 0, \tag{1.12}$$

$$\frac{d}{dt} \int_{V'} \rho' I'_{kl} dv' = 0. \tag{1.13}$$

In these equations $\rho' = \rho'(\mathbf{x}', t)$ is the density of mass and I'_{kl} is defined as

$$I'_{kl} = i'_{mn}\chi'_{mk}\chi'_{nl}$$

where i'_{mn} is the microinertia density and $\chi'_{mk} = \mathbf{e}'_m \cdot \mathbf{e}'_n$.

The conservation laws (1.12) and (1.13) can be also written in the following form:

$$\int_V \rho dv = \int_{V'} \rho' dv',$$

$$\int_V J_{mn} dv = \int_{V'} J'_{mn} dv',$$

where ρ is the density of mass and $J_{kl} = \rho I_{kl}$.

The local balance for sufficiently smooth functions can be written as

$$\rho = \rho' J,$$

$$J_{kl} = J'_{kl} J,$$

where $J = \det\left(\frac{\partial f_k}{\partial x_l}\right)$ is the Jacobian of the deformation. For the linear case

$$J \approx 1 + \frac{\partial u_k}{\partial x_k},$$

and the local form will become

$$\rho = \rho'\left(1 + \frac{\partial u_k}{\partial x_k}\right).$$

The microinertia density is conserved, which for the linear case implies that

$$I'_{kl} \approx I_{kl}.$$

1.3.5 Conservation of Linear and Angular Momentum

In the linear Cosserat Elasticity it is common to assume that the Cosserat bodies undergo small deformations. This happens when the external body forces are not exceeding certain magnitude. There are two equilibrium laws for the internal forces of a body: the conservation of linear momentum and the conservation of angular momentum. These laws are written for an arbitrary part V of the Cosserat body B that undergoes a small deformation.

The conservation of linear momentum states that

$$\int_{\partial V} \mathbf{t}^{(\mathbf{n})} ds + \int_V \rho \mathbf{b} dv = 0, \tag{1.14}$$

where $\mathbf{t}^{(\mathbf{n})}$ is the surface force per unit area acting on the surface of the body with an exterior normal $\mathbf{n}$, and $\mathbf{b}$ is the body force per unit mass [30].

The conservation of angular momentum states that

$$\int_{\partial V} \left(\mathbf{r} \times \mathbf{t}^{(\mathbf{n})} + \mathbf{m}^{(\mathbf{n})}\right) ds + \int_V \rho\left(\mathbf{r} \times \mathbf{b} + \mathbf{c}\right) dv = 0, \tag{1.15}$$

where $\mathbf{m}^{(\mathbf{n})}$ is the surface couple force per unit area acting on the surface of the body with an exterior normal $\mathbf{n}$, and $\mathbf{c}$ is the couple body force per unit mass.

If the conservation of momentum holds, then the Cauchy-Poisson Theorem [30] leads to the introduction of the concepts of the Cosserat stress tensor σ and the couple stress tensor μ:

$$\mathbf{t}^{(\mathbf{n})} = \mathbf{n} \cdot \sigma, \tag{1.16}$$

$$\mathbf{m}^{(\mathbf{n})} = \mathbf{n} \cdot \mu. \tag{1.17}$$

Substituting the expressions (1.16) and (1.17) into the equations (1.14) and (1.15) and using the divergence theorem we obtain

$$\int_V \operatorname{div} \sigma \, dv + \int_V \rho \mathbf{b} \, dv = 0,$$

and

$$\int_V (\mathbf{r} \times \operatorname{div} \sigma + \varepsilon \cdot \sigma + \operatorname{div} \mu) \, dv + \int_V \rho (\mathbf{r} \times \mathbf{b} + \mathbf{c}) \, dv = 0.$$

From the arbitrariness of the body part V, we get the static equilibrium equations in the local form:

$$\operatorname{div} \sigma + \rho \mathbf{b} = 0, \tag{1.18}$$

$$\operatorname{div} \mu + \varepsilon \cdot \sigma + \rho \mathbf{c} = 0, \tag{1.19}$$

that are usually written in the components as follows:

$$\sigma_{ji,j} + \rho b_i = 0, \tag{1.20}$$

$$\mu_{ji,j} + \varepsilon_{ijk} \sigma_{jk} + \rho c_i = 0. \tag{1.21}$$

In the absence of the body forces and couple body forces the static Cosserat equilibrium equations become

$$\sigma_{ji,j} = 0, \tag{1.22}$$

$$\mu_{ji,j} + \varepsilon_{ijk} \sigma_{jk} = 0. \tag{1.23}$$

In the dynamic case, we will follow the D'Alembert's Principle and add the inertia terms. Therefore, the equations of the Cosserat body motion take the following form:

$$\sigma_{ji,j} + \rho b_i - \rho \ddot{u}_i = 0, \tag{1.24}$$

$$\mu_{ji,j} + \varepsilon_{ijk} \sigma_{jk} + \rho c_i - J_{ij} \ddot{\varphi}_j = 0, \tag{1.25}$$

where $J_{ij} = \rho j_{ij}$.

In the absence of the body forces and couple body forces the dynamic Cosserat equilibrium equations become

$$\sigma_{ji,j} - \rho \ddot{u}_i = 0, \tag{1.26}$$

$$\mu_{ji,j} + \varepsilon_{ijk}\sigma_{jk} - J_{ij}\ddot{\varphi}_j = 0, \tag{1.27}$$

The symmetric part of the Cosserat stress tensor

$$\sigma_{\text{sym}} = \frac{1}{2}\left(\sigma + \sigma^T\right),$$

coincides with the stress tensor of the Classical Linear Elasticity. The analysis of the linearized Cosserat Elasticity, similar to the analysis of the linear Classic Elasticity made in [30], leads to the conclusion that the corresponding asymmetric Cauchy stress and couple stress tensors coincide with the asymmetric Piola-Kirchhoff stress and couple stress tensors.

1.3.6 Law of Conservation of Energy

Let E and K be the internal and kinetic energies per unit volume respectfully. The kinetic energy K is given as

$$K = \frac{1}{2}\left(\rho v_i v_j + J_{ij}\omega_i\omega_j\right), \tag{1.28}$$

where $v_i = \dot{u}$ and $\omega_i = \dot{\phi}_i$.

For the case of the isothermal deformation, the law of conservation of energy applied to any part V of the Cosserat body B can be written in the form [55]:

$$\frac{d}{dt}\int_V [E + K]\,dv = \int_V \left[\rho b_i v_i + \rho c_j \omega_i\right]dv + \int_{\partial V} [t_i v_i + m_i \omega_i]\,ds.$$

Taking into account the dynamics equilibrium equations (1.24) and (1.25) and the Cosserat strain-displacement and torsion-microrotation relations (1.10) and (1.11), we apply the divergence theorem and obtain

$$\int_V \left[\dot{E} - [\sigma_{ji}\dot{\gamma}_{ij} + \mu_{ji}\dot{\chi}_{ji}]\right]dv = 0,$$

From here we have

$$E = \sigma_{ji}\gamma_{ji} + \mu_{ji}\chi_{ji}.$$

Considering the internal energy E as a function of independent variables γ_{ji} and χ_{ji} we obtain

$$\sigma_{ji} = \frac{\partial E}{\partial \gamma_{ji}}, \qquad (1.29)$$

$$\mu_{ji} = \frac{\partial E}{\partial \chi_{ji}}. \qquad (1.30)$$

1.3.7 Cosserat Elastic Energy

Let us consider the stored strain energy of the Cosserat body B defined by the integral [55]:

$$U = \frac{1}{2} \int_B F\{\gamma, \chi\} dv, \qquad (1.31)$$

where F is the Helmholtz free energy.

We will consider isotropic, homogeneous, and central symmetric Cosserat materials. The linear approximation of the energy in a small neighborhood of $\gamma = 0$ and $\chi = 0$ can be written in the form [55]:

$$\begin{aligned}
F\{\gamma, \chi\} = {} & \frac{\mu + \alpha}{2} \gamma_{ij}\gamma_{ij} + \frac{\mu - \alpha}{2} \gamma_{ij}\gamma_{ji} + \frac{\lambda}{2}\gamma_{kk}\gamma_{nn} \\
& + \frac{\gamma + \epsilon}{2} \chi_{ij}\chi_{ij} + \frac{\gamma - \epsilon}{2} \chi_{ij}\chi_{ji} + \frac{\beta}{2}\chi_{kk}\chi_{nn},
\end{aligned}$$

where λ and μ are the Lamé parameters and α, β, γ, and ϵ are the Cosserat Elasticity parameters.

The function F is nonnegative if and only if the following conditions are satisfied [55], [34]:

$$\mu > 0,$$
$$\gamma > 0,$$
$$\alpha > 0,$$
$$\epsilon > 0,$$
$$3\lambda + 2\mu > 0,$$
$$3\beta + 2\gamma > 0,$$
$$\mu + \alpha > 0,$$
$$\gamma + \epsilon > 0.$$

For the case of the isothermal Cosserat body

$$F = E$$

and, therefore, the equations (1.29) and (1.30) will take the form:

$$\sigma_{ji} = \frac{\partial F}{\partial \gamma_{ji}},$$

$$\mu_{ji} = \frac{\partial F}{\partial \chi_{ji}}.$$

From here we obtain the constitutive formulas for the Cosserat material:

$$\sigma_{ji} = (\mu + \alpha)\gamma_{ji} + (\mu - \alpha)\gamma_{ij} + \lambda\gamma_{kk}\delta_{ij}, \tag{1.32}$$

$$\mu_{ji} = (\gamma + \epsilon)\chi_{ji} + (\gamma - \epsilon)\chi_{ij} + \beta\chi_{kk}\delta_{ij}. \tag{1.33}$$

We also consider the Cosserat body free energy [55]:

$$U_k = \int_B \Phi(\sigma, \mu)dv,$$

where

$$\Phi\{\sigma, \mu\} = \frac{\mu' + \alpha'}{2}\sigma_{ij}\sigma_{ij} + \frac{\mu' - \alpha'}{2}\sigma_{ij}\sigma_{ji} + \frac{\lambda'}{2}\sigma_{kk}\sigma_{nn}$$
$$+ \frac{\gamma' + \epsilon'}{2}\mu_{ij}\mu_{ij} + \frac{\gamma' - \epsilon'}{2}\mu_{ij}\mu_{ji} + \frac{\beta'}{2}\mu_{kk}\mu_{nn}. \tag{1.34}$$

From here we obtain the reversed constitutive formulas

$$\gamma_{ji} = \frac{\partial \Phi}{\partial \sigma_{ji}}, \tag{1.35}$$

$$\chi_{ji} = \frac{\partial \Phi}{\partial \mu_{ji}}. \tag{1.36}$$

and, therefore, the reverse form of the constitutive formulas (1.32) and (1.33) can be written as follows:

$$\gamma_{ji} = (\mu' + \alpha')\sigma_{ji} + (\mu' - \alpha')\sigma_{ij} + \lambda'\sigma_{kk}\delta_{ij}, \tag{1.37}$$

$$\chi_{ji} = (\gamma' + \epsilon')\mu_{ji} + (\gamma' - \epsilon')\mu_{ij} + \beta'\mu_{kk}\delta_{ij}, \tag{1.38}$$

where

$$\mu' = \frac{1}{4\mu},$$

$$\alpha' = \frac{1}{4\alpha},$$

$$\gamma' = \frac{1}{4\gamma},$$

$$\epsilon' = \frac{1}{4\epsilon},$$

$$\lambda' = -\frac{\lambda}{2\mu(3\lambda + 2\mu)},$$

$$\beta' = -\frac{\beta}{2\gamma(3\beta + 2\gamma)}.$$

1.4 VARIATIONAL PRINCIPLE FOR COSSERAT ELASTOSTATICS

Let $\mathcal{A}$ be the set of all kinematically admissible states $\mathfrak{s} = [\mathbf{u}, \phi, \sigma, \mu, \gamma, \chi]$ that satisfy the strain-displacement and torsion-microrotation relations (1.10) and (1.11). Let the functional $\Theta(\mathfrak{s})$ be defined as follows:

$$\Theta(\mathfrak{s}) = U - \int_B [\sigma \cdot \gamma + \mu \cdot \chi] dv$$

$$+ \int_{\mathcal{G}_1} \left[\hat{\sigma}_\mathbf{n} \cdot (\mathbf{u} - \hat{\mathbf{u}}) + \hat{\mu}_\mathbf{n} \left(\phi - \hat{\phi} \right) \right] ds$$

$$+ \int_{\mathcal{G}_2} [\hat{\sigma} \cdot \mathbf{u} + \hat{\mu} \cdot \phi] ds.$$

The Variational Principle for Cosserat Elastostatics states that the zero variation

$$\delta\Theta(\mathfrak{s}) = 0,$$

at a kinematically admissible state $\mathfrak{s} \in \mathcal{A}$ is equivalent to $\mathfrak{s}$ being a solution of the system of equilibrium equations (1.20) and (1.21), which satisfies the constitutive formulas (1.32) and (1.33) and the mixed boundary conditions

$$\mathbf{u} = \hat{\mathbf{u}} \text{ on } \mathcal{G}_1, \tag{1.39}$$

$$\phi = \hat{\phi} \text{ on } \mathcal{G}_1, \tag{1.40}$$

$$\sigma_\mathbf{n} = \hat{\sigma} \text{ on } \mathcal{G}_2, \tag{1.41}$$

$$\mu_\mathbf{n} = \hat{\mu} \text{ on } \mathcal{G}_2. \tag{1.42}$$

Proof of the Variational Principle for Cosserat Elastostatics

Let us consider the variation of the functional $\Theta(\mathfrak{s})$:

$$\delta\Theta(\mathfrak{s}) = \delta U(S) - \int_B [\delta\sigma\cdot\gamma + \sigma\cdot\delta\gamma + \delta\mu\cdot\chi + \mu\cdot\delta\chi]\,dv$$

$$+ \int_{\mathcal{G}_1} [\delta\sigma_\mathbf{n}\cdot(\mathbf{u}-\hat{\mathbf{u}}) + \sigma_\mathbf{n}\delta\mathbf{u} + \delta\mu_\mathbf{n}\cdot(\phi-\hat{\phi}) + \mu_\mathbf{n}\delta\phi]\,ds$$

$$+ \int_{\mathcal{G}_2} [\hat{\sigma}\cdot\delta\mathbf{u} + \hat{\mu}\cdot\delta\phi]\,ds.$$

After the integrating by parts we have

$$\int_B \sigma\cdot\delta\gamma\,dv = \int_{\partial B} \sigma_\mathbf{n}\cdot\delta\mathbf{u}\,ds - \int_B \delta\mathbf{u}\cdot\operatorname{div}\sigma\,dv + \int_B \varepsilon\cdot\sigma\cdot\delta\phi\,dv,$$

$$\int_B \mu\cdot\delta\chi\,dv = \int_{\partial B} \mu_\mathbf{n}\cdot\delta\phi\,ds - \int_B \delta\phi\cdot\operatorname{div}\mu\,dv.$$

From the equation

$$\delta F = \frac{\partial\Phi}{\partial\sigma}\cdot\delta\sigma + \frac{\partial\Phi}{\partial\mu}\cdot\delta\mu,$$

we obtain

$$\delta\Theta(\mathfrak{s}) = \int_B \left[\left(\frac{\partial\Phi}{\partial\sigma}-\gamma\right)\cdot\delta\sigma\right]dv + \int_B \left[\left(\frac{\partial\Phi}{\partial\mu}-\chi\right)\cdot\delta\mu\right]dv$$

$$+ \int_B [\operatorname{div}\sigma\cdot\delta u]\,dv + \int_B [(\operatorname{div}\mu+\varepsilon\cdot\sigma)\cdot\delta\mathbf{u}]\,dv$$

$$+ \int_{\mathcal{G}_1} [(\mathbf{u}-\hat{\mathbf{u}})\cdot\delta\sigma_\mathbf{n}]\,ds + \int_{\mathcal{G}_1} \left[\left(\phi-\hat{\phi}\right)\cdot\delta\mu_\mathbf{n}\right]ds$$

$$+ \int_{\mathcal{G}_2} [(\sigma_\mathbf{n}-\hat{\sigma})\cdot\delta\mathbf{u}]\,ds + \int_{\mathcal{G}_2} [(\mu_\mathbf{n}-\hat{\mu})\cdot\delta\phi]\,ds.$$

Therefore, $\delta\Theta(\mathfrak{s}) = 0$ if the admissible state $\mathfrak{s}$ is a solution of the Cosserat Elasticity problem:

$$\text{div } \sigma = 0,$$
$$\text{div } \mu + \varepsilon \cdot \sigma = 0,$$

that satisfies the equations

$$\frac{\partial \Phi}{\partial \sigma} = \gamma,$$
$$\frac{\partial \Phi}{\partial \mu} = \chi,$$

and the boundary conditions (4.5) – (4.8).

Summary of Cosserat Linear Elastostatics

Equilibrium Equations

$\sigma_{ji,j} = 0$

$\mu_{ji,j} + \varepsilon_{ijk}\sigma_{jk} = 0$

Strain-Displacement and Torsion-Microrotation Relations

$\gamma_{ji} = u_{i,j} + \varepsilon_{ijk}\phi_k$

$\chi_{ji} = \phi_{i,j}$

Constitutive Formulas

$\sigma_{ji} = (\mu + \alpha)\gamma_{ji} + (\mu - \alpha)\gamma_{ij} + \lambda\gamma_{kk}\delta_{ij}$

$\mu_{ji} = (\gamma + \epsilon)\chi_{ji} + (\gamma - \epsilon)\chi_{ij} + \beta\chi_{kk}\delta_{ij}$

1.5 PRINCIPLE OF MINIMUM POTENTIAL ENERGY

For simplicity, the constructive formulas (1.32) and (1.33) can be written in the following short form:

$$\sigma = A_\gamma[\gamma], \tag{1.43}$$
$$\mu = A_\chi[\chi], \tag{1.44}$$

where A_γ and A_χ are linear operators.

Let $[\mathbf{u}, \phi, \gamma, \chi, \sigma, \mu]$ be an elastic state of the Cosserat body corresponding to a body force $\mathbf{b}$ and a couple force $\mathbf{c}$. In other words, this elastic state satisfies the equilibrium equations (1.18) and (1.19), the constitutive formulas (1.32) and (1.33), and the strain-displacement and torsion-microrotation relations (1.10) and (1.11). Therefore, we have following equality

$$\int_{\partial B} [\sigma_{\mathbf{n}} \cdot \mathbf{u} + \mu_{\mathbf{n}} \cdot \phi]\, ds + \int_B [\mathbf{b} \cdot \mathbf{u} + \mathbf{c} \cdot \phi]\, dv = U(\gamma, \chi), \qquad (1.45)$$

where the strain energy $U(\gamma, \chi)$ is given as

$$U(\gamma, \chi) = \frac{1}{2} \int_B \Big(A_\gamma[\gamma] \cdot \gamma + A_\chi[\chi] \cdot \chi\Big) dv,$$

Indeed, we have

$$\int_{\partial B} [\sigma_{\mathbf{n}} \cdot \mathbf{u} + \mu_{\mathbf{n}} \cdot \phi]\, ds = \int_{\partial B} [(\sigma \cdot \mathbf{u})\mathbf{n} + (\mu \cdot \phi)\mathbf{n}]\, ds$$

$$= \int_B [\mathrm{div}\,(\sigma \cdot \mathbf{u}) + \mathrm{div}\,(\mu \cdot \phi)]\, dv$$

$$= \int_B [\mathbf{u} \cdot \mathrm{div}\,\sigma + \sigma \cdot \nabla\mathbf{u} + \phi \cdot \mathrm{div}\,\mu + \mu \cdot \nabla\phi]\, dv$$

$$= -\int_B [\rho\mathbf{b} \cdot \mathbf{u} + \rho\mathbf{c} \cdot \phi]\, dv + \int_B [\sigma \cdot \gamma + \mu \cdot \chi]\, dv.$$

and, therefore, the equality (1.45) holds.

Now we will formulate the Principle of Minimum Potential Energy for Cosserat Elasticity.

Principle of Minimum Potential Energy for Cosserat Elasticity

Let $\mathcal{A}$ be the set of all kinematically admissible states $\mathfrak{s} = [\mathbf{u}, \phi, \sigma, \mu, \gamma, \chi]$ that satisfy the strain-displacement and torsion-microrotation relations (1.10) and (1.11). Let the functional $\Pi(\mathfrak{s})$ be defined as follows:

$$\Pi(\mathfrak{s}) = U - \int_{\mathcal{G}_2} [\sigma \cdot \mathbf{u} + \mu \cdot \phi]\, ds. \qquad (1.46)$$

The Principle of Minimum Potential Energy for Cosserat Elasticity states that

$$\Pi(\mathfrak{s}) \le \Pi(\tilde{\mathfrak{s}}),$$

where $\tilde{\mathfrak{s}} = \left[\tilde{\mathbf{u}}, \tilde{\phi}, \tilde{\sigma}, \tilde{\mu}, \tilde{\gamma}, \tilde{\chi}\right] \in \mathcal{A}$ is any kinematically admissible state and $\mathfrak{s}$ is a solution of the system of equilibrium equations (1.20) and (1.21), which satisfies the constitutive formulas (1.32) and (1.33) and the mixed boundary conditions

$$\mathbf{u} = \hat{\mathbf{u}} \text{ on } \mathcal{G}_1, \tag{1.47}$$

$$\phi = \hat{\phi} \text{ on } \mathcal{G}_1, \tag{1.48}$$

$$\sigma_{\mathbf{n}} = \hat{\sigma} \text{ on } \mathcal{G}_2, \tag{1.49}$$

$$\mu_{\mathbf{n}} = \hat{\mu} \text{ on } \mathcal{G}_2. \tag{1.50}$$

The equality holds if and only if

$$\tilde{\mathbf{u}} - \mathbf{u} = \bar{\mathbf{u}},$$

$$\tilde{\phi} - \phi = \bar{\phi},$$

where $\bar{\mathbf{u}}$ is an infinitesimal rigid displacement and $\bar{\phi}$ is an infinitesimal rigid microrotation of the Cosserat body B.

Proof of the Principle of Minimum Potential Energy for Cosserat Elasticity

Since the solutions are exact and kinematically admissible then we can introduce the variables

$$\bar{\gamma} = \tilde{\gamma} - \gamma = (\nabla\bar{\mathbf{u}})^T - \boldsymbol{\epsilon} \cdot \bar{\phi},$$

$$\bar{\chi} = \tilde{\chi} - \chi = \left(\nabla\bar{\phi}\right)^T,$$

where $\bar{\mathbf{u}}$ and $\bar{\phi}$ satisfy the boundary conditions

$$\bar{\mathbf{u}} = 0 \text{ on } \mathcal{G}_1,$$

$$\bar{\phi} = 0 \text{ on } \mathcal{G}_1.$$

Then we have

$$\begin{aligned}
\tilde{\gamma} \cdot A_\gamma[\tilde{\gamma}] + \tilde{\chi} \cdot A_\chi[\tilde{\chi}] = {} & \gamma \cdot A_\gamma[\gamma] + \chi \cdot A_\chi[\chi] \\
& + \bar{\gamma} \cdot A_\gamma[\bar{\gamma}] + \bar{\chi} \cdot A_\chi[\bar{\chi}] \\
& + \gamma \cdot A_\gamma[\bar{\gamma}] + \chi \cdot A_\chi[\bar{\chi}] \\
& + \bar{\gamma} \cdot A_\gamma[\gamma] + \bar{\chi} \cdot A_\chi[\chi].
\end{aligned}$$

The last expression can be simplified as follows:

$$\tilde{\gamma} \cdot A_\gamma [\tilde{\gamma}] + \tilde{\chi} \cdot A_\chi [\tilde{\chi}] = \gamma \cdot A_\gamma [\gamma] + \chi \cdot A_\chi [\chi]$$
$$+ \bar{\gamma} \cdot A_\gamma [\bar{\gamma}] + \bar{\chi} \cdot A_\chi [\bar{\chi}]$$
$$+ \sigma \cdot \bar{\gamma} + \mu \cdot \bar{\chi}.$$

The latter expression implies that

$$U(\tilde{\gamma}, \tilde{\chi}) - U(\gamma, \chi) = U(\bar{\gamma}, \bar{\chi}) + \int_B [\sigma \cdot \bar{\gamma} + \mu \cdot \bar{\chi}] dv. \qquad (1.51)$$

Using the divergence theorem together with the equation (1.45) we obtain

$$\int_B [\sigma \cdot \bar{\gamma} + \mu \cdot \bar{\chi}] dv = \int_{\mathcal{G}_2} \left[\sigma_\mathbf{n} \cdot \bar{\mathbf{u}} + \mu_\mathbf{n} \cdot \bar{\phi} \right] ds \qquad (1.52)$$

$$+ \int_B \left[\rho \mathbf{b} \cdot \bar{\mathbf{u}} + \rho \mathbf{c} \cdot \bar{\phi} \right] dv. \qquad (1.53)$$

From here we have

$$\int_B [\sigma \cdot \bar{\gamma} + \mu \cdot \bar{\chi}] dv = \int_{\mathcal{G}_2} \left[\hat{\sigma} \cdot \bar{\mathbf{u}} + \hat{\mu} \cdot \bar{\phi} \right] ds \qquad (1.54)$$

$$+ \int_B \left[\rho \mathbf{b} \cdot \bar{\mathbf{u}} + \rho \mathbf{c} \cdot \bar{\phi} \right] dv. \qquad (1.55)$$

Taking into account the definition of the functional (1.46) and the formulas (1.51)–(1.55) we obtain

$$\Pi(\tilde{\mathfrak{s}}) - \Pi(\mathfrak{s}) = U(\bar{\gamma}, \bar{\chi})$$

Since the energy $U(\bar{\gamma}, \bar{\chi})$ is nonnegative we conclude that

$$\Pi(\tilde{\mathfrak{s}}) \geq \Pi(\mathfrak{s})$$

and the equality occurs if and only if

$$\bar{\gamma} = 0,$$
$$\bar{\chi} = 0,$$

i.e., when $\bar{\mathbf{u}}$ is an infinitesimal rigid displacement and $\bar{\phi}$ is an infinitesimal rigid microrotation.

1.6 VARIATIONAL PRINCIPLE FOR COSSERAT ELASTODYNAMICS

Let $\mathcal{A}$ be the set of all kinematically admissible states $\mathfrak{s} = [\mathbf{u}, \phi, \sigma, \mu, \gamma, \chi]$ that satisfy the strain-displacement and torsion-microrotation relations (1.10) and (1.11). Let the functional $\Theta(\mathfrak{s})$ be defined as follows:

$$
\Theta(\mathfrak{s}) = \int_{t_0}^{t_1} [U - T]\, dt
$$

$$
- \int_{t_0}^{t_1} \int_B \left[\sigma \cdot \gamma + \mu \cdot \chi - \mathbf{p} \cdot \frac{\partial \mathbf{u}}{\partial t} - \mathbf{q} \cdot \frac{\partial \phi}{\partial t} \right] dv\, dt
$$

$$
+ \int_B [(\mathbf{p} - \mathbf{p}_0) \cdot (\mathbf{u} - \mathbf{u}_0) + (\mathbf{q} - \mathbf{q}_0) \cdot (\phi - \phi_0)]\, dv
$$

$$
+ \int_{t_0}^{t_1} \int_{\mathcal{G}_1} \left[\sigma_{\mathbf{n}} \cdot (\mathbf{u} - \hat{\mathbf{u}}) + \mu_{\mathbf{n}} \cdot \left(\phi - \hat{\phi} \right) \right] ds\, dt
$$

$$
+ \int_{t_0}^{t_1} \int_{\mathcal{G}_2} [\hat{\sigma} \cdot \mathbf{u} + \hat{\mu} \cdot \phi]\, ds\, dt.
$$

The Variational Principle for Cosserat Elastodynamics states that the zero variation

$$
\delta \Theta(\mathfrak{s}) = 0,
$$

at a kinematically admissible state $\mathfrak{s} \in \mathcal{A}$ is equivalent to $\mathfrak{s}$ being a solution of the system of equilibrium equations (1.24) and (1.25), which satisfies the constitutive formulas (1.32) and (1.33), the mixed boundary conditions

$$
\mathbf{u} = \hat{\mathbf{u}} \text{ on } \mathcal{G}_1, \tag{1.56}
$$

$$
\phi = \hat{\phi} \text{ on } \mathcal{G}_1, \tag{1.57}
$$

$$
\sigma_{\mathbf{n}} = \hat{\sigma} \text{ on } \mathcal{G}_2, \tag{1.58}
$$

$$
\mu_{\mathbf{n}} = \hat{\mu} \text{ on } \mathcal{G}_2. \tag{1.59}
$$

and the initial conditions

$$\mathbf{u}(t_0) = \mathbf{u_0}, \tag{1.60}$$

$$\phi(t_0) = \phi_0, \tag{1.61}$$

$$\mathbf{p}(t_0) = \mathbf{p_0}, \tag{1.62}$$

$$\mathbf{q}(t_0) = \mathbf{q_0}. \tag{1.63}$$

Proof of the Variational Principle for Cosserat Elastodynamics

Let us consider the variation of the functional $\Theta(\mathfrak{s})$:

$$\delta\Theta(\mathfrak{s}) = \int_{t_0}^{t_1} (\delta U - \delta T)\, dt$$

$$- \int_{t_0}^{t_1} \int_B [\delta\boldsymbol{\sigma}\cdot\boldsymbol{\gamma} + \boldsymbol{\sigma}\cdot\delta\boldsymbol{\gamma} + \delta\boldsymbol{\mu}\cdot\boldsymbol{\chi} + \boldsymbol{\mu}\cdot\delta\boldsymbol{\chi}]\, dv\, dt$$

$$+ \int_{t_0}^{t_1} \int_B \left[\delta\mathbf{p}\cdot\frac{\partial\mathbf{u}}{\partial t} + \mathbf{p}\cdot\delta\left(\frac{\partial\mathbf{u}}{\partial t}\right) + \delta\mathbf{q}\cdot\frac{\partial\phi}{\partial t} + \mathbf{q}\cdot\delta\left(\frac{\partial\phi}{\partial t}\right)\right] dv\, dt$$

$$- \int_B [\delta\mathbf{p}\cdot(\mathbf{u}-\mathbf{u_0}) + \delta\mathbf{q}\cdot(\phi-\phi_0)]\, dv\, dt$$

$$- \int_B [(\mathbf{p}-\mathbf{p_0})\cdot\delta\mathbf{u} + (\mathbf{q}-\mathbf{q_0})\cdot\delta\phi]\, dv\, dt$$

$$+ \int_{t_0}^{t_1} \int_{\mathcal{G}_1} \left[\delta\boldsymbol{\sigma_n}\cdot(\mathbf{u}-\hat{\mathbf{u}}) + \boldsymbol{\sigma_n}\cdot\delta\mathbf{u} + \delta\boldsymbol{\mu_n}\cdot(\phi-\hat{\phi}) + \boldsymbol{\mu_n}\cdot\delta\phi\right] ds\, dt$$

$$+ \int_{t_0}^{t_1} \int_{\mathcal{G}_2} [\hat{\boldsymbol{\sigma}}\cdot\delta\mathbf{u} + \hat{\boldsymbol{\mu}}\cdot\delta\phi]\, ds\, dt.$$

Now we integrate by parts:

$$\int_B \boldsymbol{\sigma}\cdot\delta\boldsymbol{\gamma}\, dv = \int_{\partial B} \boldsymbol{\sigma_n}\cdot\delta\mathbf{u}\, ds - \int_B \delta\mathbf{u}\cdot\operatorname{div}\boldsymbol{\sigma}\, dv + \int_B \boldsymbol{\varepsilon}\cdot\boldsymbol{\sigma}\cdot\delta\phi\, dv,$$

$$\int_B \boldsymbol{\mu}\cdot\delta\boldsymbol{\chi}\, dv = \int_{\partial B} \boldsymbol{\mu_n}\cdot\delta\phi\, ds - \int_B \delta\phi\cdot\operatorname{div}\boldsymbol{\mu}\, dv.$$

We also have

$$\int_{t_0}^{t_1} \int_B \left[\delta\left(\frac{\partial \mathbf{u}}{\partial t}\right) \cdot \mathbf{p} + \delta\left(\frac{\partial \phi}{\partial t}\right) \cdot \mathbf{q} \right] dv\, dt = \int_B [\delta\mathbf{u} \cdot (\mathbf{p} - \mathbf{p}(t_0))]\, dv$$

$$+ \int_B [\delta\phi \cdot (\mathbf{q} - \mathbf{q}(t_0))]\, dv$$

$$- \int_{t_0}^{t_1} \int_B \left[\delta\frac{\partial \mathbf{p}}{\partial t} \cdot \mathbf{u} + \delta\phi \cdot \frac{\partial \mathbf{q}}{\partial t} \right] dv\, dt,$$

and

$$\int_{t_0}^{t_1} \int_B \left[\frac{\partial \mathbf{u}}{\partial t} \cdot \delta\mathbf{p} + \delta\mathbf{q} \cdot \frac{\partial \phi}{\partial t} \right] dv\, dt = \int_B [\delta\mathbf{p} \cdot (\mathbf{u} - \mathbf{u}(t_0))]\, dv$$

$$+ \int_B [\delta\mathbf{q} \cdot (\phi - \phi(t_0))]\, dv$$

$$- \int_{t_0}^{t_1} \int_B \left[\delta\frac{\partial \mathbf{p}}{\partial t} \cdot \mathbf{u} + \delta\phi \cdot \frac{\partial \mathbf{q}}{\partial t} \right] dv\, dt.$$

Using the equations

$$\delta F = \frac{\partial \Phi}{\partial \sigma} \cdot \delta\sigma + \frac{\partial \Phi}{\partial \mu} \cdot \delta\mu,$$

$$\delta K = \frac{\partial K}{\partial \mathbf{p}} \cdot \delta\mathbf{p} + \frac{\partial K}{\partial \mathbf{q}} \cdot \delta\mathbf{q},$$

we can rewrite the expression for the variation $\delta\Theta(\mathfrak{s})$ in the following form:

$$\delta\Theta(\mathfrak{s}) = \int_{t_0}^{t_1} \int_B \left[\left(\frac{\partial \Phi}{\partial \sigma} - \gamma\right) \cdot \delta\sigma \right] dv\, dt$$

$$+ \int_{t_0}^{t_1} \int_B \left[\left(\frac{\partial \Phi}{\partial \mu} - \chi\right) \cdot \delta\mu \right] dv\, dt$$

$$-\int_{t_0}^{t_1}\int_B \left[\left(\frac{\partial K}{\partial \mathbf{p}}-\frac{\partial \mathbf{u}}{\partial t}\right)\cdot \delta\mathbf{p}\right]dvdt$$

$$-\int_{t_0}^{t_1}\int_B \left[\left(\frac{\partial K}{\partial \mathbf{q}}-\frac{\partial \phi}{\partial t}\right)\cdot \delta\mathbf{q}\right]dvdt$$

$$+\int_{t_0}^{t_1}\int_B \left[\left(\operatorname{div}\boldsymbol{\sigma}-\frac{\partial \mathbf{p}}{\partial t}\right)\cdot \delta\mathbf{u}\right]dvdt$$

$$+\int_{t_0}^{t_1}\int_B \left[\left(\operatorname{div}\boldsymbol{\mu}+\boldsymbol{\varepsilon}\cdot\boldsymbol{\sigma}-\frac{\partial \mathbf{q}}{\partial t}\right)\cdot \delta\mathbf{u}\right]dvdt$$

$$+\int_B \left[\delta\mathbf{p}\cdot(\mathbf{u}-\mathbf{u}(t_0))+\delta\mathbf{q}\cdot(\phi-\phi(t_0))\right]dvdt$$

$$+\int_B \left[(\mathbf{p}-\mathbf{p}(t_0))\cdot \delta\mathbf{u}+(\mathbf{q}-\mathbf{q}(t_0))\cdot \delta\phi\right]dvdt$$

$$-\int_B \left[\delta\mathbf{p}\cdot(\mathbf{u}-\mathbf{u}_0)+\delta\mathbf{q}\cdot(\phi-\phi_0)\right]dvdt$$

$$-\int_B \left[(\mathbf{p}-\mathbf{p}_0)\cdot \delta\mathbf{u}+(\mathbf{q}-\mathbf{q}_0)\cdot \delta\phi\right]dvdt$$

$$+\int_{t_0}^{t_1}\int_{\mathcal{G}_1} \left[(\mathbf{u}-\hat{\mathbf{u}})\cdot \delta\boldsymbol{\sigma}_\mathbf{n}\right]dsdt$$

$$+\int_{t_0}^{t_1}\int_{\mathcal{G}_1} \left[\left(\phi-\hat{\phi}\right)\cdot \delta\boldsymbol{\mu}_\mathbf{n}\right]dsdt$$

$$+\int_{t_0}^{t_1}\int_{\mathcal{G}_2} \left[(\boldsymbol{\sigma}_\mathbf{n}-\hat{\boldsymbol{\sigma}})\cdot \delta\mathbf{u}\right]dsdt$$

$$+\int_{t_0}^{t_1}\int_{\mathcal{G}_2} \left[(\boldsymbol{\mu}_\mathbf{n}-\hat{\boldsymbol{\mu}})\cdot \delta\phi\right]dsdt.$$

Therefore, $\delta\Theta(\mathfrak{s}) = 0$ if the admissible state $\mathfrak{s}$ is the solution of the Cosserat Elasticity problem

$$\operatorname{div} \boldsymbol{\sigma} = \frac{\partial \mathbf{p}}{\partial t},$$

$$\operatorname{div} \boldsymbol{\mu} + \boldsymbol{\varepsilon} \cdot \boldsymbol{\sigma} = \frac{\partial \mathbf{q}}{\partial t},$$

that satisfies the equations

$$\frac{\partial \Phi}{\partial \boldsymbol{\sigma}} = \gamma,$$

$$\frac{\partial \Phi}{\partial \boldsymbol{\mu}} = \chi,$$

$$\rho \frac{\partial \mathbf{u}}{\partial t} = \mathbf{p},$$

$$J \frac{\partial \varphi}{\partial t} = \mathbf{q},$$

the boundary conditions (1.56)–(1.59) and the initial conditions (1.60)–(1.63).

Summary of Cosserat Linear Elastodynamics

Equilibrium Equations

$\sigma_{ji,j} - \rho \ddot{u}_i = 0$

$\mu_{ji,j} + \varepsilon_{ijk}\sigma_{jk} - J_{ij}\ddot{\phi}_j = 0$

Strain-Displacement and Torsion-Microrotation Relations

$\gamma_{ji} = u_{i,j} + \varepsilon_{ijk}\phi_k$

$\chi_{ji} = \phi_{i,j}$

Constitutive Formulas

$\sigma_{ji} = (\mu + \alpha)\gamma_{ji} + (\mu - \alpha)\gamma_{ij} + \lambda\gamma_{kk}\delta_{ij}$

$\mu_{ji} = (\gamma + \epsilon)\chi_{ji} + (\gamma - \epsilon)\chi_{ij} + \beta\chi_{kk}\delta_{ij}$

CHAPTER 2

Analysis of Cosserat Sample Bodies

2.1 SCOPE OF THE CHAPTER

This chapter provides the exact solutions for the three-dimensional Cosserat Elasticity. The solutions are derived for the Cosserat rectangular cuboids of different width-to-thickness ratios. The obtained data is used for the validation of the Cosserat Plate Theory and the Finite Element Method for Cosserat Plates. It also provides the insight into the geometric restrictions for the Cosserat plates.

2.2 COSSERAT SAMPLE BODIES

Let us consider the Cosserat body B being a rectangular cuboid $[0,a] \times [0,a] \times \left[-\frac{h}{2}, \frac{h}{2}\right]$. Let the sets T and B be the top and the bottom surfaces contained in the planes $x_3 = \frac{h}{2}$ and $x_3 = -\frac{h}{2}$ respectively, and the curve $\Gamma = \Gamma_1 \cup \Gamma_2$ be the lateral part of the boundary:

$$\Gamma_1 = \left\{ (x_1, x_2, x_3) : x_1 \in \{0, a\}, x_2 \in [0, a], x_3 \in \left[-\frac{h}{2}, \frac{h}{2}\right] \right\},$$

$$\Gamma_2 = \left\{ (x_1, x_2, x_3) : x_1 \in [0, a], x_2 \in \{0, a\}, x_3 \in \left[-\frac{h}{2}, \frac{h}{2}\right] \right\}.$$

DOI: 10.1201/9781003190264-2

We will use the following technical parameters [36], [44]:

Young's modulus $\qquad E = \dfrac{\mu(3\lambda + 2\mu)}{\lambda + \mu}$

Poisson's ratio $\qquad v = \dfrac{\lambda}{2(\lambda + \mu)}$

Characteristic length for bending $\qquad l_b = \dfrac{1}{2}\sqrt{\dfrac{\gamma + \epsilon}{\mu}}$

Characteristic length for torsion $\qquad l_t = \sqrt{\dfrac{\gamma}{\mu}}$

Coupling number $\qquad N = \sqrt{\dfrac{\alpha}{\mu + \alpha}}$

Taking into account that the ratio β/γ is equal to 1 for bending [45], we can convert the values of the technical parameters to the values of Lamé and Cosserat Elasticity parameters using the following formulas:

$$\lambda = \frac{Ev}{2v^2 + v - 1},$$
$$\mu = \frac{E}{2(v+1)},$$
$$\alpha = \frac{EN^2}{2(v+1)(N^2 - 1)},$$
$$\beta = \frac{El_t}{2(v+1)},$$
$$\gamma = \frac{El_t^2}{2(v+1)},$$
$$\epsilon = \frac{E\left(4l_b^2 - l_t^2\right)}{2(v+1)}.$$

In our computations we consider the Cosserat bodies made of polyurethane foam – a material reported in the literature to behave like a Cosserat material. We will use the values of the technical elastic parameters presented in [45]:

$$E = 299.5 \text{ MPa},$$
$$v = 0.44,$$
$$l_b = 0.327 \text{ mm},$$
$$l_t = 0.62 \text{ mm},$$
$$N^2 = 0.04.$$

The values of the technical constants correspond to the following values of Lamé and Cosserat Elasticity parameters:

$$\lambda = 762.616 \text{ MPa},$$
$$\mu = 103.993 \text{ MPa},$$
$$\alpha = 4.333 \text{ MPa},$$
$$\beta = 39.975 \text{ MPa},$$
$$\gamma = 39.975 \text{MPa},$$
$$\epsilon = 4.505 \text{ MPa}.$$

We consider a low density rigid foam usually characterized by the densities of 24–50 kg/m^3 [50]. In the further computations we use the density value $\rho = 34$ kg/m^3 and values of the microinertia $J_x = 0.001$, $J_y = 0.001$, $J_z = 0.001$ for the Cosserat square body of thickness $h = 0.1$ and the width-to-thickness ratio a/h in the interval [5, 30].

2.3 STATICS OF COSSERAT SAMPLE BODIES

We consider the Cosserat body with the boundary conditions given in Table 2.1. We study the deformation of the body based on the three-dimensional Cosserat equilibrium equations without body forces and body moments:

$$\sigma_{ji,j} = 0, \tag{2.1}$$
$$\mu_{ji,j} + \varepsilon_{ijk}\sigma_{jk} = 0, \tag{2.2}$$

TABLE 2.1 Boundary conditions for simply supported Cosserat body.

Boundary Γ_1	Boundary Γ_2	Boundary T	Boundary B
$u_2 = 0$	$u_1 = 0$	$\sigma_{33} = p$	$\sigma_{33} = 0$
$u_3 = 0$	$u_3 = 0$	$\mu_{33} = 0$	$\mu_{33} = 0$
$\phi_1 = 0$	$\phi_2 = 0$		
$\sigma_{11} = 0$	$\sigma_{22} = 0$		
$\mu_{12} = 0$	$\mu_{21} = 0$		
$\mu_{13} = 0$	$\mu_{23} = 0$		

accompanied by the constitutive formulas (1.32) and (1.33) and strain-displacement and torsion-microrotation relations (1.10) and (1.11), where the initial distribution of the pressure is given as

$$p(x_1, x_2) = \sin\left(\frac{\pi x_1}{a}\right) \cdot \sin\left(\frac{\pi x_2}{a}\right).$$

Using the separation of variables and taking into account the boundary conditions given in Table 2.1, we express the kinematic variables in the form:

$$u_1 = \cos\left(\frac{\pi x_1}{a}\right) \cdot \sin\left(\frac{\pi x_2}{a}\right) \cdot z_1(x_3), \tag{2.3}$$

$$u_2 = \sin\left(\frac{\pi x_1}{a}\right) \cdot \cos\left(\frac{\pi x_2}{a}\right) \cdot z_1(x_3), \tag{2.4}$$

$$u_3 = \sin\left(\frac{\pi x_1}{a}\right) \cdot \sin\left(\frac{\pi x_2}{a}\right) \cdot z_2(x_3), \tag{2.5}$$

$$\phi_1 = \sin\left(\frac{\pi x_1}{a}\right) \cdot \cos\left(\frac{\pi x_2}{a}\right) \cdot z_3(x_3), \tag{2.6}$$

$$\phi_2 = -\cos\left(\frac{\pi x_1}{a}\right) \cdot \sin\left(\frac{\pi x_2}{a}\right) \cdot z_3(x_3), \tag{2.7}$$

$$\phi_3 = 0, \tag{2.8}$$

where the functions $z_1(x_3)$, $z_2(x_3)$, and $z_3(x_3)$ represent the transverse variations of the kinematic variables.

If we substitute the expressions (2.3)–(2.8) into (1.10) and (1.11), then into (1.32) and (1.33) and then into (2.1) and (2.2), we will obtain the following second order linear system of three ordinary differential equations in terms of z_1, z_2 and z_3:

$$a^2(\mu+\alpha)z_1'' + a(\mu-\alpha+\lambda)z_2' + 2a^2\alpha z_3' - 2\pi^2(\lambda+2\mu)z_1 = 0, \tag{2.9}$$

$$a^2(\lambda+2\mu)z_2'' - 2a\pi(\mu-\alpha+\lambda)z_1' - 2\pi(\mu+\alpha)z_2 + 4a\pi z_3 = 0, \tag{2.10}$$

$$a^2(\epsilon+\gamma)z_3'' - 2a^2\alpha z_1' + 2a\alpha\pi z_2 - 2\left(2a^2\alpha+(\epsilon+\gamma)\pi^2\right)z_3 = 0, \tag{2.11}$$

complemented by the boundary conditions on $x_3 = \frac{h}{2}$:

$$a(\mu+\alpha)z_1' + (\mu-\alpha)\pi z_2 = 0,$$

$$a(\lambda+2\mu)z_2' - 2\pi\lambda z_1 = a,$$

$$z_3 = 0,$$

and the boundary conditions on $x_3 = -\frac{h}{2}$:

$$a(\mu + \alpha)z_1' + (\mu - \alpha)\pi z_2 = 0,$$
$$a(\lambda + 2\mu)z_2' - 2\pi\lambda z_1 = 0,$$
$$z_3 = 0.$$

Now let us make the following substitutions:

$$\kappa_1(x_3) = z_1(x_3),$$
$$\kappa_2(x_3) = z_2(x_3),$$
$$\kappa_3(x_3) = z_3(x_3),$$
$$\kappa_4(x_3) = z_1'(x_3),$$
$$\kappa_5(x_3) = z_2'(x_3),$$
$$\kappa_6(x_3) = z_3'(x_3).$$

The system of equations (2.9)–(2.11) can be, therefore, rewritten as a first order linear system of six ordinary differential equations in terms of κ_i $(i = 1, ..., 6)$:

$$\boldsymbol{\kappa}'(x_3) = \mathbf{K} \cdot \boldsymbol{\kappa}(x_3), \tag{2.12}$$

or equivalently in the matrix form:

$$
\begin{bmatrix} \kappa_1' \\ \kappa_2' \\ \kappa_3' \\ \kappa_4' \\ \kappa_5' \\ \kappa_6' \end{bmatrix} =
\begin{bmatrix}
0 & 0 & 0 & 1 & 0 & 0 \\
0 & 0 & 0 & 0 & 1 & 0 \\
0 & 0 & 0 & 0 & 0 & 1 \\
K_{41} & 0 & 0 & 0 & K_{45} & K_{46} \\
0 & K_{52} & K_{53} & K_{54} & 0 & 0 \\
0 & K_{62} & K_{63} & K_{64} & 0 & 0
\end{bmatrix}
\begin{bmatrix} \kappa_1 \\ \kappa_2 \\ \kappa_3 \\ \kappa_4 \\ \kappa_5 \\ \kappa_6 \end{bmatrix},
$$

where

$$K_{41} = \frac{2\pi^2(\lambda + 2\mu)}{a^2(\alpha + \mu)},$$

$$K_{45} = \frac{\pi(\alpha - \lambda - \mu)}{a(\alpha + \mu)},$$

$$K_{46} = -\frac{2a}{(\alpha + \mu)},$$

$$K_{52} = \frac{2\pi^2(\alpha + \mu)}{a^2(\lambda + 2\mu)},$$

$$K_{53} = -\frac{4\pi\alpha}{a(\lambda + 2\mu)},$$

$$K_{54} = \frac{2\pi(-\alpha + \lambda + \mu)}{a(\lambda + 2\mu)},$$

$$K_{62} = -\frac{2\pi\alpha}{a(\gamma + \epsilon)},$$

$$K_{63} = \frac{2(2a^2\alpha + \pi^2(\gamma + \epsilon))}{a^2(\gamma + \epsilon)},$$

$$K_{64} = \frac{2\alpha}{\gamma + \epsilon}.$$

The system is complemented by the boundary conditions at $x_3 = \pm\frac{h}{2}$:

$$a(\mu + \alpha)\kappa_4 + (\mu - \alpha)\pi\kappa_2 = 0, \qquad (2.13)$$

$$a(\lambda + 2\mu)\kappa_5 - 2\pi\lambda\kappa_1 = a, \qquad (2.14)$$

$$\kappa_3 = 0. \qquad (2.15)$$

The general solution of the system of six ordinary differential equations given above is found as follows:

$$\kappa(x_3) = c_i e^{r_i x_3} \theta_i^T, \qquad (2.16)$$

where r_i are the eigenvalues of the matrix $\mathbf{K}$, θ_i are the corresponding eigenvectors of the matrix $\mathbf{K}$ and c_i are some constants that can be determined from the boundary conditions (2.13)–(2.15). Note that since $z_i = \kappa_i$ for $i = 1, 2, 3$, then the general solution for the variables z_i is also of the form (2.16) (see [24], [47] for details).

The eigenvalues r_i of the matrix $\mathbf{K}$ are

$$r_1 = -\frac{\sqrt{2}\pi}{a},$$

$$r_2 = \frac{\sqrt{2}\pi}{a},$$

$$r_3 = -\sqrt{\frac{4\alpha\mu}{(\gamma + \epsilon)(\alpha + \mu)} + \frac{2\pi^2}{a^2}},$$

$$r_4 = \sqrt{\frac{4\alpha\mu}{(\gamma + \epsilon)(\alpha + \mu)} + \frac{2\pi^2}{a^2}}.$$

The corresponding eigenvectors θ_i of the matrix **K** are

$$\theta_1 = \left[\frac{a}{2\pi}, -\frac{a}{\sqrt{2\pi}}, 0, -\frac{1}{\sqrt{2}}, 1, 0\right],$$

$$\theta_2 = \left[\frac{a}{2\pi}, \frac{a}{\sqrt{2\pi}}, 0, \frac{1}{\sqrt{2}}, 1, 0\right],$$

$$\theta_3 = \left[-\frac{\gamma+\epsilon}{2\mu}, \frac{\pi(\gamma+\epsilon)}{a\mu r_4}, -\frac{1}{r_4}, \frac{(\gamma+\epsilon)r_4}{\mu}, -\frac{\pi(\gamma+\epsilon)}{a\mu}, 1\right],$$

$$\theta_4 = \left[-\frac{\gamma+\epsilon}{2\mu}, -\frac{\pi(\gamma+\epsilon)}{a\mu r_4}, \frac{1}{r_4}, -\frac{(\gamma+\epsilon)r_4}{\mu}, -\frac{\pi(\gamma+\epsilon)}{a\mu}, 1\right].$$

The numerical values of the maxima and minima of the kinematic variables u_i and ϕ_i for different values of the width-to-thickness ratios a/h are given in Tables 2.2 and 2.3. The distribution of the displacements u_i, microrotations ϕ_i, stresses σ_{ij}, and couple stresses μ_{ij} for the thickness-to-width ratios $a/h = 30$ are given in Figures 2.1–2.4. The graphs show a linear type of distribution for the displacements u_α and stresses $\sigma_{\alpha\beta}$ and the quadratic type of distributions for the displacement u_3, and stresses $\sigma_{\alpha 3}$ and $\sigma_{3\beta}$. The graphs also show a constant type of distribution for the microrotations ϕ_α and couple stresses $\mu_{\alpha\beta}$.

TABLE 2.2 Extrema of the kinematic variables u_i and ϕ_i calculated using the three-dimensional Cosserat Elasticity for the width-to-thickness ratio $a/h = 5, 10, 15$.

a/h	5	10	15
u_1	0.00008525	0.00062141	0.00204337
u_2	0.00008525	0.00062141	0.00204337
u_3	0.00374336	0.02374431	0.06242824
ϕ_1	0.00009767	0.00133873	0.00526100
ϕ_2	−0.00009767	−0.00133873	−0.00526100
ϕ_3	0.00000000	0.00000000	0.00000000

TABLE 2.3   Extrema of the kinematic variables u_i and ϕ_i calculated using the three-dimensional Cosserat Elasticity for the width-to-thickness ratio $a/h = 20, 25, 30$.

a/h	20	25	30
u_1	0.00481496	0.00935158	0.01610936
u_2	0.00481496	0.00935158	0.01610936
u_3	0.12076264	0.20133599	0.30787077
ϕ_1	0.01325449	0.02664900	0.04679946
ϕ_2	−0.01325449	−0.02664900	−0.04679946
ϕ_3	0.00000000	0.00000000	0.00000000

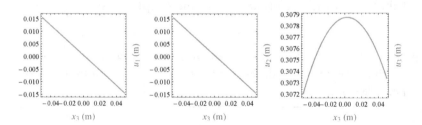

Figure 2.1   Distributions of the displacements u_i for the Cosserat body with the width-to-thickness ratio for $a/h = 30$.

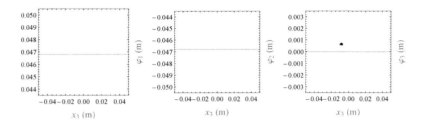

Figure 2.2 Distributions of the microrotations ϕ_i for the Cosserat body with the width-to-thickness ratio for $a/h = 30$.

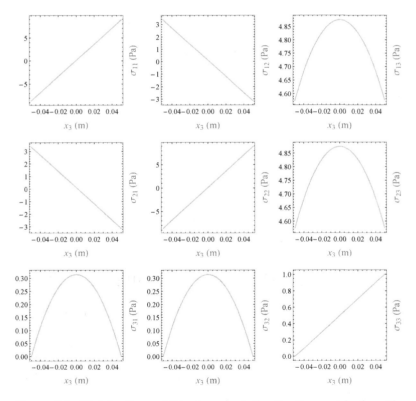

Figure 2.3 Distributions of the stresses σ_{ij} for the Cosserat body with the width-to-thickness ratio for $a/h = 30$.

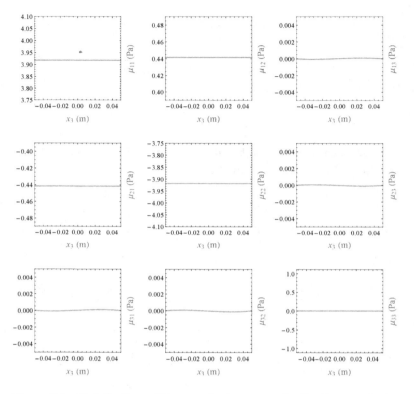

Figure 2.4 Distributions of the couple stresses μ_{ij} for the Cosserat body with the width-to-thickness ratio for $a/h = 30$.

2.4 DYNAMICS OF COSSERAT SAMPLE BODIES

We solve the three-dimensional Cosserat equilibrium equations without body forces:

$$\sigma_{ji,j} - \rho \ddot{u}_i = 0, \tag{2.17}$$

$$\mu_{ji,j} + \varepsilon_{ijk}\sigma_{jk} - J_{ij}\ddot{\varphi}_j = 0, \tag{2.18}$$

accompanied by the constitutive equations (1.32) and (1.33) and strain-displacement and torsion-microrotation relations (1.10) and (1.11), complemented by the boundary conditions given in Table 2.4.

TABLE 2.4   Boundary conditions for simply supported Cosserat body.

Boundary Γ_1	Boundary Γ_2	Boundary T	Boundary B
$u_2 = 0$	$u_1 = 0$	$\sigma_{33} = p$	$\sigma_{33} = 0$
$u_3 = 0$	$u_3 = 0$	$\mu_{33} = 0$	$\mu_{33} = 0$
$\phi_1 = 0$	$\phi_2 = 0$		
$\sigma_{11} = 0$	$\sigma_{22} = 0$		
$\mu_{12} = 0$	$\mu_{21} = 0$		
$\mu_{13} = 0$	$\mu_{23} = 0$		

The initial distribution of the pressure is assume to be

$$p = \sin\left(\frac{\pi x_1}{a}\right) \cdot \sin\left(\frac{\pi x_2}{a}\right) \cdot \sin \omega t.$$

The microinertia tensor $\mathbf{J}$ is assumed to have a diagonal form

$$\mathbf{J} = \begin{bmatrix} J_x & 0 & 0 \\ 0 & J_y & 0 \\ 0 & 0 & J_z \end{bmatrix}. \tag{2.19}$$

Using the method of separation of variables and taking into account the boundary conditions given in Table 2.4, we express the kinematic variables in the form:

$$u_1 = \cos\left(\frac{\pi x_1}{a}\right) \cdot \sin\left(\frac{\pi x_2}{a}\right) \cdot z_1(x_3) \cdot \sin \omega t, \tag{2.20}$$

$$u_2 = \sin\left(\frac{\pi x_1}{a}\right) \cdot \cos\left(\frac{\pi x_2}{a}\right) \cdot z_2(x_3) \cdot \sin \omega t, \tag{2.21}$$

$$u_3 = \sin\left(\frac{\pi x_1}{a}\right) \cdot \sin\left(\frac{\pi x_2}{a}\right) \cdot z_3(x_3) \cdot \sin \omega t, \tag{2.22}$$

$$\phi_1 = \sin\left(\frac{\pi x_1}{a}\right) \cdot \cos\left(\frac{\pi x_2}{a}\right) \cdot z_4(x_3) \cdot \sin \omega t, \tag{2.23}$$

$$\phi_2 = \cos\left(\frac{\pi x_1}{a}\right) \cdot \sin\left(\frac{\pi x_2}{a}\right) \cdot z_5(x_3) \cdot \sin \omega t, \tag{2.24}$$

$$\phi_3 = \cos\left(\frac{\pi x_1}{a}\right) \cdot \cos\left(\frac{\pi x_2}{a}\right) \cdot z_6(x_3) \cdot \sin \omega t, \tag{2.25}$$

where the functions $z_i(x_3)$ represent the transverse variations of the kinematic variables.

If we substitute the expressions (2.3)–(2.8) into (1.10) and (1.11), then into (1.32) and (1.33) and then into (2.17) and (2.18), we will obtain the following eigenvalue problem

$$\mathbf{B}z = \omega^2 \mathbf{A}z, \qquad (2.26)$$

where

$$\mathbf{B} = \begin{bmatrix} L_3 & b_3 L_0 & b_4 L_1 & 0 & -b_5 L_1 & -b_6 L_0 \\ b_3 L_0 & L_3 & b_4 L_1 & b_5 L_1 & 0 & b_6 L_0 \\ -b_4 L_1 & b_4 L_1 & b_7 L_2 & b_6 L_0 & -b_6 L_0 & 0 \\ 0 & -b_5 L_1 & b_6 L_0 & L_4 & b_{11} L_0 & b_{12} L_1 \\ b_5 L_1 & 0 & -b_6 L_0 & b_{11} L_0 & L_4 & b_{12} L_1 \\ -b_6 L_0 & b_6 L_0 & 0 & -b_{12} L_1 & -b_{12} L_1 & L_5 \end{bmatrix},$$

$$\mathbf{A} = \begin{bmatrix} -a^2\rho & 0 & 0 & 0 & 0 & 0 \\ 0 & -a^2\rho & 0 & 0 & 0 & 0 \\ 0 & 0 & -a^2\rho & 0 & 0 & 0 \\ 0 & 0 & 0 & -a^2 J_x & 0 & 0 \\ 0 & 0 & 0 & 0 & -a^2 J_y & 0 \\ 0 & 0 & 0 & 0 & 0 & -a^2 J_z \end{bmatrix},$$

$$z = \begin{bmatrix} z_1, & z_2, & z_3, & z_4, & z_5, & z_6 \end{bmatrix}^T,$$

The differential operators L_i are defined as follows:

$$L_0 = I,$$

$$L_1 = \frac{d}{dx_3},$$

$$L_2 = \frac{d^2}{dx_3^2},$$

$$L_3 = b_1 L_2 + b_2 L_0,$$

$$L_4 = b_9 L_2 + b_{10} L_0,$$

$$L_5 = b_{13} L_2 + b_2 L_0,$$

and the coefficients b_i are given as

$$b_1 = a^2(\mu + \alpha), \qquad\qquad b_2 = -\pi^2(\alpha + \lambda + 3\mu),$$
$$b_3 = -\pi^2(\lambda + \mu - \alpha), \qquad b_4 = a\pi(\lambda + \mu - \alpha),$$
$$b_5 = 2a^2\alpha, \qquad\qquad\qquad b_6 = 2a\pi\alpha,$$
$$b_7 = a^2(2\mu + \lambda), \qquad\qquad b_8 = -2\pi^2(\alpha + \mu),$$
$$b_9 = a^2(\gamma + \epsilon), \qquad\qquad b_{10} = -\pi^2(\beta + \epsilon + 3\gamma),$$
$$b_{11} = -\pi^2(\beta + \gamma - \epsilon), \qquad b_{12} = -a\pi(\beta + \gamma - \epsilon),$$
$$b_{13} = a^2(\beta + 2\gamma), \qquad\qquad b_{14} = -2\pi^2(\gamma + \epsilon) - 4a^2\alpha.$$

The system of differential equations (2.26) is complemented by the boundary conditions for $x_3 = \frac{h}{2}$

$$\mathbf{D}z = D_0,$$

and the boundary conditions for $x_3 = -\frac{h}{2}$

$$\mathbf{D}z = 0.$$

The operator $\mathbf{D}$ is given as

$$\mathbf{D} = \begin{bmatrix} d_1 L_1 & 0 & d_2 L_0 & 0 & -d_3 L_0 & 0 \\ 0 & d_1 L_1 & d_2 L_0 & d_3 L_0 & 0 & 0 \\ d_4 L_0 & d_4 L_0 & d_5 L_1 & 0 & 0 & 0 \\ 0 & 0 & 0 & d_6 L_1 & 0 & d_7 L_0 \\ 0 & 0 & 0 & 0 & d_6 L_1 & d_7 L_0 \\ 0 & 0 & 0 & d_8 L_0 & d_8 L_0 & d_9 L_1 \end{bmatrix}.$$

The vector D_0 is given as

$$D_0 = \begin{bmatrix} 0, & 0, & a, & 0, & 0, & 0 \end{bmatrix}^T.$$

The coefficients d_i are defined as

$$d_1 = a(\mu + \alpha), \qquad\qquad d_2 = -\pi(\mu - \alpha),$$
$$d_3 = 2a\alpha, \qquad\qquad\qquad d_4 = a(\lambda + 2\mu),$$
$$d_5 = -\pi\lambda, \qquad\qquad\qquad d_6 = a(\gamma + \epsilon),$$
$$d_7 = a(\gamma - \epsilon), \qquad\qquad d_8 = \pi\beta,$$
$$d_9 = a(\beta + 2\gamma).$$

TABLE 2.5 Resonance frequencies calculated using the three-dimensional Cosserat Elasticity for the values of the width-to-thickness ratio $a/h = 5, 10, 15$.

a/h	5	10	15
Kirchhoff frequency ω_K (Hz)	8.96958	3.56063	2.18625
Mindlin-Reissner frequency ω_{MR} (Hz)	64.9537	58.7313	57.4417
Cosserat frequency ω_{C1} (Hz)	1679.21	1007.43	671.602
Cosserat frequency ω_{C2} (Hz)	3021.39	1342.89	1007.19

We solve the system (2.26) using the Wolfram Mathematica software. When the amplitude of the solutions starts to grow indefinitely, the oscillation of the Cosserat body corresponds to its resonance frequency. Tables 2.5 and 2.6 show the resonance frequencies of a Cosserat body calculated using the three-dimensional Cosserat Elasticity.

TABLE 2.6 Resonance frequencies calculated using the three-dimensional Cosserat Elasticity for the values of the width-to-thickness ratio $a/h = 20, 25, 30$.

a/h	20	25	30
Kirchhoff frequency ω_K (Hz)	1.56808	1.21281	0.97999
Mindlin-Reissner frequency ω_{MR} (Hz)	56.9537	56.6994	56.5352
Cosserat frequency ω_{C1} (Hz)	504.798	392.006	335.805
Cosserat frequency ω_{C2} (Hz)	729.005	616.850	504.704

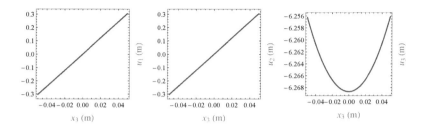

Figure 2.5 Distributions of the displacements u_i for the Cosserat sample body vibrating on the frequency of 1 Hz, with the value of the width-to-thickness ratio $a/h = 30$.

2.5 PLATE GEOMETRY RESTRICTION FOR COSSERAT BODIES

For the values of the width-to-thickness ratio a/h of at least 5 the distributions for the displacements u_α and stresses $\sigma_{\alpha\beta}$ are uniformly linear and the distributions for the displacements u_3 and the stresses $\sigma_{\alpha 3}$ and $\sigma_{3\beta}$ are uniformly quadratic. However, when the values of the width-to-thickness ratio a/h less than 5, the distribution of the displacements u_α and stresses $\sigma_{\alpha\beta}$ stop being linear and the distribution for the displacement u_3 stops being a symmetric parabola. This implies that there must be a restriction for the width-to-thickness ratio, which in case of the square Cosserat plate-like body should have the value of at least 5. Figures 2.5–2.8 show the distributions of the stresses and

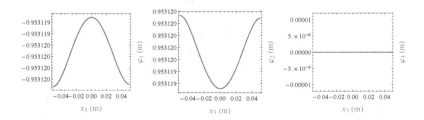

Figure 2.6 Distributions of the microrotations ϕ_i for the Cosserat sample body vibrating on the frequency of 1 Hz, with the value of the width-to-thickness ratio $a/h = 30$.

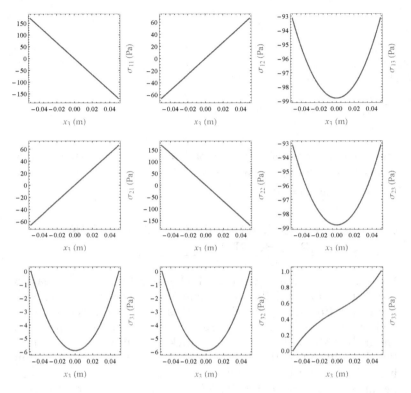

Figure 2.7 Distributions of the stresses σ_{ij} for the Cosserat sample body vibrating on the frequency of 1 Hz, with the value of the width-to-thickness ratio $a/h = 30$.

kinematic variables of a Cosserat sample body vibrating on the frequency of 1 Hz. Figures 2.9–2.12 show the distributions of the stresses and kinematic variables of a Cosserat sample body with the value of the width-to-thickness ratio $a/h = 3$.

For simple shapes (circles, ellipses, squares, rectangles, etc.) we can estimate the "representative length" of the plate-like body and impose the restrictions on the representative-length-to-thickness ratio. Let A be the area of the cross section P. Let d be the diameter of P, i.e. be the largest distance between any two points in P. We can estimate the representative length L for the

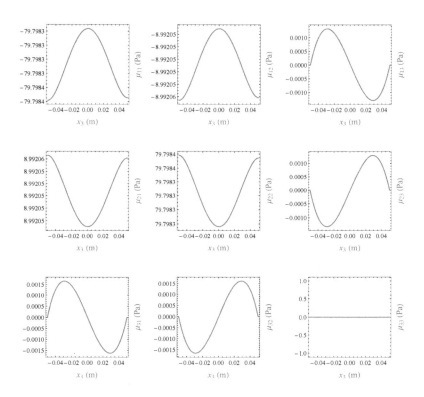

Figure 2.8 Distributions of the couple stresses μ_{ij} for the Cosserat sample body vibrating on the frequency of 1 Hz, with the value of the width-to-thickness ratio $a/h = 30$.

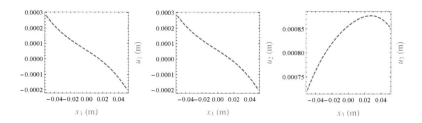

Figure 2.9 Distributions of the displacements u_i for the Cosserat body with the value of the width-to-thickness ratio $a/h = 3$.

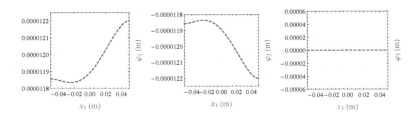

Figure 2.10   Distributions of the microrotations ϕ_i for the Cosserat body with the value of the width-to-thickness ratio $a/h = 3$.

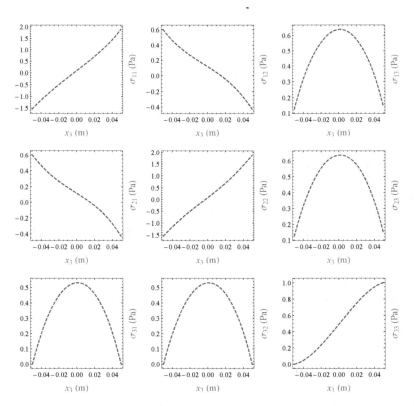

Figure 2.11   Distributions of the stresses σ_{ij} for the Cosserat body with the value of the width-to-thickness ratio $a/h = 3$.

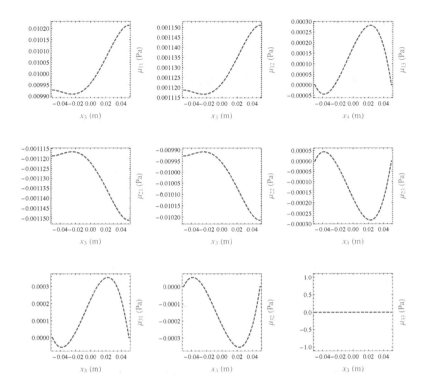

Figure 2.12 Distributions of the couple stresses μ_{ij} for the Cosserat body with the value of the width-to-thickness ratio $a/h = 3$.

plate-like body:

$$L = c \cdot \frac{A}{d}, \qquad (2.27)$$

where $c = \frac{4}{\pi}$.

For simple shapes of the body B, it can be considered to be a plate if the following restriction for the representative-length-to-thickness ratio is satisfied:

$$5 \le \frac{L}{h} \le 100. \qquad (2.28)$$

TABLE 2.7 Representative-length-to-thickness ratio L/h and diameter-to-representative-length ratio d/L for plates of simple shapes.

Plate Shape	Diameter	Area	L/h	d/L
Circle of radius r	$2r$	πr^2	$\frac{2r}{h}$	1
Ellipse of semi-axes a and b $(a > b)$	$2a$	πab	$\frac{2b}{h}$	$\frac{a}{b}$
Square of side a	a^2	$a\sqrt{2}$	$\frac{2\sqrt{2}a}{\pi h}$	$\frac{\pi}{2}$
Rectangle of sides a and b $(a > b)$	$\sqrt{a^2 + b^2}$	ab	$\frac{4ab}{\sqrt{a^2+b^2}h\pi}$	$\frac{(a^2+b^2)\pi}{4ab}$

To distinguish the plates from the beams we will also require the diameter-to-representative-length ratio d/L to satisfy:

$$1 \le \frac{d}{L} \le 10. \tag{2.29}$$

Table 2.7 presents the values of the representative-length-to-thickness ratio L/h and the diameter-to-representative-length ratio d/L for plates for some simple shapes. The classification of the Cosserat plates depending on the representative length to thickness ratio L/h is provided in Table 2.8.

In the case of the square plates of side a, the plates with the width-to-thickness ratio a/h between 5 and 10 are considered to be thick, a/h between 10 and 20 are considered to be moderately thick, a/h between 20 and 50 are considered to be thin, and a/h between 50 and 100 are considered to be very thin. The classification of the square Cosserat plates depending on the width-to-thickness ratio a/h is provided in Table 2.9 width-to-thickness ratio.

In the case of the circular plates of diameter d, the plates with the diameter-to-thickness ratio d/h between 4 and 9 are

TABLE 2.8   Classification of Cosserat plates depending on the representative-length-to-thickness ratio L/h.

Representative-Length-to-Thickness Ratio L/h	Type of Plate
$4 \leq L/h \leq 9$	Thick Plate
$9 < L/h \leq 18$	Moderately Thick Plate
$18 < L/h \leq 45$	Thin Plate
$45 < L/h \leq 90$	Very Thin Plate

considered to be thick, d/h between 9 and 18 are considered to be moderately thick, d/h between 18 and 45 are considered to be thin, and d/h between 45 and 90 are considered to be very thin. The classification of the circular Cosserat plates depending on the diameter-to-thickness ratio d/h is provided in Table 2.10 diameter-to-thickness ratio.

TABLE 2.9   Classification of square Cosserat plates depending on the width-to-thickness ratio a/h.

Width-to-Thickness Ratio a/h	Type of Plate
$5 \leq a/h \leq 10$	Thick Plate
$10 < a/h \leq 20$	Moderately Thick Plate
$20 < a/h \leq 50$	Thin Plate
$50 < a/h \leq 100$	Very Thin Plate

TABLE 2.10   Classification of circular Cosserat plates depending on the diameter-to-thickness ratio d/h.

Diameter-to-Thickness Ratio d/h	Type of Plate
$4 \leq d/h \leq 9$	Thick Plate
$9 < d/h \leq 18$	Moderately Thick Plate
$18 < d/h \leq 45$	Thin Plate
$45 < d/h \leq 90$	Very Thin Plate

The Cosserat Plate Theory presented in this book is consistent with the three-dimensional Cosserat Elasticity for the thick, moderately thick, thin, and very thin plates.

CHAPTER 3

Cosserat Plate Statics

3.1 SCOPE OF THE CHAPTER

This chapter introduces the static Theory of Cosserat Plates. We assume the distributions of the Cosserat stress, couple stress, displacement, and microrotation. The approximations are consistent with the three-dimensional Cosserat Elasticity. We obtain the Cosserat plate equilibrium equations, constitutive formulas, strain-displacement and torsion-microrotation relations. The system of equilibrium equations is parametric, where the optimal value of the parameter minimizes the Cosserat elastic energy. The introduction of the optimal parameter makes the Cosserat plate model well posed. We also present the elliptic system of Cosserat plate field equations written in terms of the kinematic variables.

3.2 PLATE ASSUMPTIONS

3.2.1 Boundary Conditions

Let B be the three-dimensional Cosserat body and ∂B be its boundary. We split the boundary ∂B into $\mathcal{G}_u$ and $\mathcal{G}_\sigma$, such that $\mathcal{G}_u \cup \mathcal{G}_\sigma = \partial B$. We prescribe the displacements and microrotations on $\mathcal{G}_u$ and the stress and couple stress on $\mathcal{G}_\sigma$. Therefore, the equilibrium equations (1.22) and (1.23), the strain-displacement and torsion-microrotation relations (1.10) and (1.11) and the

DOI: 10.1201/9781003190264-3

47

constitutive equations (1.32) and (1.33) are considered together with the following boundary conditions prescribed on ∂B:

$$u_i = \hat{u}_i \text{ on } \mathcal{G}_u, \tag{3.1}$$

$$\phi_i = \hat{\phi}_i \text{ on } \mathcal{G}_u, \tag{3.2}$$

$$\sigma_{ji} n_i = \hat{\sigma}_j \text{ on } \mathcal{G}_\sigma, \tag{3.3}$$

$$\mu_{ji} n_i = \hat{\mu}_j \text{ on } \mathcal{G}_\sigma, \tag{3.4}$$

where n_i denotes the outward unit normal vector to ∂B.

We consider the thin plate P where h is the thickness of the plate and $x_3 = 0$ contains its middle plane. The sets T and B are the top and the bottom surfaces contained in the planes $x_3 = h/2$ and $x_3 = -h/2$, respectively. The curve Γ is the boundary of the middle plane of the plate.

The set of points $P = \left(\Gamma \times \left[-\frac{h}{2}, \frac{h}{2} \right] \right) \cup T \cup B$ forms the entire surface of the plate. We split the boundary of the middle plane of the plate Γ into Γ_u and Γ_σ, such that $\Gamma_u \cup \Gamma_\sigma = \Gamma$. The set of points $\Gamma_u \times \left[-\frac{h}{2}, \frac{h}{2} \right]$ represents the lateral part of the boundary where displacements and microrotations are prescribed. The set of points $\Gamma_\sigma \times \left[-\frac{h}{2}, \frac{h}{2} \right]$ represents the lateral part of the boundary where stress and couple stress are prescribed. For simplicity, we will also use the notation P for the middle plane internal domain of the plate.

Let us consider the vertical load boundary conditions at the top and bottom of the plate, which can be written in the form:

$$\sigma_{33}(x_1, x_2, h/2) = \sigma^t(x_1, x_2), \tag{3.5}$$

$$\sigma_{33}(x_1, x_2, -h/2) = \sigma^b(x_1, x_2), \tag{3.6}$$

$$\sigma_{3\beta}(x_1, x_2, h/2) = 0, \tag{3.7}$$

$$\sigma_{3\beta}(x_1, x_2, -h/2) = 0, \tag{3.8}$$

$$\mu_{33}(x_1, x_2, h/2) = 0, \tag{3.9}$$

$$\mu_{33}(x_1, x_2, -h/2) = 0, \tag{3.10}$$

$$\mu_{3\beta}(x_1, x_2, h/2) = 0, \tag{3.11}$$

$$\mu_{3\beta}(x_1, x_2, -h/2) = 0, \tag{3.12}$$

where $(x_1, x_2) \in P$.

3.2.2 Stress Approximations

The starting point of our consideration is the three-dimensional Cosserat Elastostatics, which includes the equilibrium equations, the strain-displacement and torsion-microrotation relations, and the constitutive formulas. The Cosserat Plate Theory is two-dimensional; it is based on approximations of the plate stresses and displacements by polynomials with respect x_3 variable, which have to be consistent with three-dimensional Cosserat theory.

We assume that the variation of the stress σ_{ji} and couple stress μ_{ji} components across the thickness can be represented as polynomials of the variable x_3 in such way that it will be consistent with the equilibrium equations, which in the absence of the body forces and body moments become

$$\sigma_{ji,j} = 0, \tag{3.13}$$

$$\varepsilon_{ijk}\sigma_{jk} + \mu_{ji,j} = 0. \tag{3.14}$$

In all further formulas, when the greek letters α and β are used for the indices, we will assume that $\alpha, \beta \in \{1, 2\}$. We will also use the scaled vertical component ζ defined as

$$\zeta = \frac{2}{h}x_3.$$

First, as it is assumed in the classical theory of plates, we use the expressions for the stress components $\sigma_{\alpha\beta}$ in the following form [11]:

$$\sigma_{\alpha\beta} = m_{\alpha\beta}(x_1, x_2)\zeta, \tag{3.15}$$

Note that the first two differential equations of the stress equilibrium (3.13) can be written as

$$\sigma_{j\beta,j} = 0.$$

Using the first two equations of the stress equilibrium (3.13) and the assumption (3.15) we obtain the shear stress components $\sigma_{3\beta}$ in the following form:

$$\sigma_{3\beta} = q_\beta(x_1, x_2)\left(1 - \zeta^2\right). \tag{3.16}$$

We will use the expressions for the stress components $\sigma_{3\beta}$ similar to the expressions for the components $\sigma_{\alpha 3}$ with an additional constant term [27]:

$$\sigma_{\alpha 3} = q_\alpha^*(x_1, x_2)\left(1 - \zeta^2\right) + \hat{q}_\alpha(x_1, x_2). \tag{3.17}$$

If we substitute the equations (3.17) into the third equation of the stress equilibrium (3.13):

$$\sigma_{j3,j} = 0,$$

we will obtain the cubic form for the transverse normal stress σ_{33}

$$\sigma_{33} = k^*(x_1, x_2)\left(\zeta - \frac{\zeta^3}{3}\right) + l^*(x_1, x_2)\zeta + m^*(x_1, x_2). \tag{3.18}$$

The next step is to accommodate approximations (3.18) with the boundary conditions (3.5) and (3.6). By direct substitution it easy to obtain that

$$\sigma_{33} = \frac{3}{4}\left(\zeta - \frac{\zeta^3}{3}\right)p^*(x_1, x_2) + \frac{1}{2}\hat{p}(x_1, x_2)\zeta + \sigma_0(x_1, x_2), \tag{3.19}$$

where

$$p^*(x_1, x_2) + \hat{p}(x_1, x_2) = p(x_1, x_2) \tag{3.20}$$

We consider the parametric solution of the equation (3.20) in the following form

$$p^*(x_1, x_2) = \eta p(x_1, x_2), \tag{3.21}$$
$$\hat{p}(x_1, x_2) = (1 - \eta)p(x_1, x_2), \tag{3.22}$$

where η is a parameter, which we will call the **splitting parameter**. This parameter allows us to split the bending pressure on the plate $p(x_1, x_2)$ into two parts corresponding to different orders of stress approximation. The optimal value of the splitting parameter shows the contribution of different types of approximation in the plate bending. This approach gives a more accurate description of the mechanical phenomenon of bending.

From here we have the expressions for $p(x_1, x_2)$ and $\sigma_0(x_1, x_2)$:

$$p(x_1, x_2) = \sigma^t(x_1, x_2) - \sigma^b(x_1, x_2) \tag{3.23}$$

$$\sigma_0(x_1, x_2) = \frac{1}{2}\left(\sigma^t(x_1, x_2) + \sigma^b(x_1, x_2)\right) \tag{3.24}$$

Note that the form (3.19) will automatically satisfy the boundary condition requirements (3.5) and (3.6). For the case of $\eta = 1$ the expression (3.19) is identical to the expression of σ_{33} given in [26].

The stress assumptions (3.15), (3.16), (3.17), (3.18), and (3.19) are confirmed by the exact solution of the Cosserat sample plate.

3.2.3   Couple Stress Approximations

We use the following approximations for the couple stress components $\mu_{\alpha\beta}$:

$$\mu_{\alpha\beta} = r^*_{\alpha\beta}(x_1, x_2)\left(1 - \zeta^2\right) + \hat{r}_{\alpha\beta}(x_1, x_2), \tag{3.25}$$

and couple stress components $\mu_{\alpha3}$:

$$\mu_{\alpha3} = s_\alpha(x_1, x_2)\zeta. \tag{3.26}$$

By substituting the couple stress (3.25) into the first two differential equations of the equilibrium of angular momentum (3.14):

$$\varepsilon_{\beta jk}\sigma_{jk} + \mu_{j\beta,j} = 0,$$

and taking into account (3.15) and (3.16), we obtain the expression for the transverse shear couple stress as

$$\mu_{3\beta} = s^*_\beta(x_1, x_2)\left(\zeta - \frac{\zeta^3}{3}\right). \tag{3.27}$$

Substituting (3.27) into the boundary conditions (3.11) and (3.12) we obtain that

$$s^*_\beta(x_1, x_2) = 0,$$

i.e. the transverse shear couple stress $\mu_{3\beta}$ is equal to zero:

$$\mu_{3\beta} = 0. \tag{3.28}$$

We will also use the following assumption that μ_{33} is zero:

$$\mu_{33} = 0 \tag{3.29}$$

which automatically satisfies the boundary conditions (3.9) and (3.10).

The couple stress assumptions (3.25), (3.26), (3.28), and (3.29) are confirmed by the exact solution of the Cosserat sample plate.

3.2.4   Kinematic Approximations

From the strain-displacement relation

$$\gamma_{ji} = u_{i,j} + \varepsilon_{ijk}\phi_k,$$

we have that

$$\gamma_{\alpha\alpha} = u_{\alpha,\alpha},$$

and the expressions for u_α are similar to $\gamma_{\alpha\alpha}$. From the first two equations of the constitutive formula

$$\sigma_{ji} = (\mu + \alpha)\gamma_{ji} + (\mu - \alpha)\gamma_{ij} + \lambda\gamma_{kk}\delta_{ij},$$

we have that the expression for $\gamma_{\alpha\alpha}$ is similar to the expression for $\sigma_{\alpha\alpha}$, which are linear in terms of ζ. Therefore, we choose the components u_α to also be linear in terms of ζ

$$u_\alpha = \psi_\alpha(x_1, x_2)\zeta. \tag{3.30}$$

From the torsion-microrotation relation

$$\chi_{ji} = \phi_{i,j},$$

we have that

$$\chi_{\alpha\alpha} = \phi_{\alpha,\alpha},$$

and the expressions for ϕ_α are similar to $\chi_{\alpha\alpha}$. From the first two equations of the constitutive formula

$$\mu_{ji} = (\gamma + \epsilon)\chi_{ji} + (\gamma - \epsilon)\chi_{ij} + \beta\chi_{kk}\delta_{ij},$$

we have that the expression for $\chi_{\alpha\alpha}$ is similar to the expression for $\mu_{\alpha\alpha}$. Therefore, we choose the components ϕ_α to be in the form similar to $\chi_{\alpha\alpha}$

$$\phi_\alpha = \omega_\alpha^*(x_1, x_2)\left(1 - \zeta^2\right) + \hat{\omega}_\alpha(x_1, x_2). \tag{3.31}$$

From the strain-displacement relation

$$\gamma_{ji} = u_{i,j} + \varepsilon_{ijk}\phi_k,$$

we also have that

$$\gamma_{13} = u_{3,1} + \phi_2,$$
$$\gamma_{23} = u_{3,2} - \phi_1.$$

This implies that the expression for the component u_3 is similar to the the expression for the $\gamma_{\alpha 3}$. From the third equation of the constitutive formula

$$\sigma_{ji} = (\mu + \alpha)\gamma_{ji} + (\mu - \alpha)\gamma_{ij} + \lambda\gamma_{kk}\delta_{ij},$$

we have that the expression for $\gamma_{3\alpha}$ is similar to the expression for $\sigma_{3\beta}$ and $\sigma_{\alpha 3}$. Therefore, we choose the component u_3 to be in the form

$$u_3 = w^*(x_1, x_2)\left(1 - \zeta^2\right) + \hat{w}(x_1, x_2). \tag{3.32}$$

From the torsion-microrotation relation

$$\chi_{ji} = \phi_{i,j},$$

we also have that

$$\chi_{\alpha 3} = \phi_{3,\alpha},$$

This implies that the expression for the component ϕ_3 is similar to the the expression for the $\chi_{\alpha 3}$. From the third equation of the constitutive formula

$$\mu_{ji} = (\gamma + \epsilon)\chi_{ji} + (\gamma - \epsilon)\chi_{ij} + \beta\chi_{kk}\delta_{ij},$$

we have that the expression for $\chi_{3\alpha}$ is similar to the expression for $\mu_{\alpha 3}$ and $\mu_{3\beta}$. Therefore, we choose the component ϕ_3 to be in the following form:

$$\phi_3 = \omega_3(x_1, x_2)\zeta, \tag{3.33}$$

The kinematic assumptions (3.30), (3.31), (3.32), and (3.33) are confirmed by the exact solution of the Cosserat sample plate.

Summary of the Cosserat Plate Assumptions

Stress assumptions

$$\sigma_{\alpha\beta} = m_{\alpha\beta}(x_1, x_2)\zeta$$

$$\sigma_{3\beta} = q_\beta(x_1, x_2)\left(1 - \zeta^2\right)$$

$$\sigma_{\alpha3} = q_\alpha^*(x_1, x_2)\left(1 - \zeta^2\right) + \hat{q}_\alpha(x_1, x_2)$$

$$\sigma_{33} = \frac{3}{4}\left(\zeta - \frac{\zeta^3}{3}\right)p^*(x_1, x_2) + \frac{1}{2}\hat{p}(x_1, x_2)\zeta + \sigma_0(x_1, x_2)$$

Couple stress assumptions

$$\mu_{\alpha\beta} = r_{\alpha\beta}^*(x_1, x_2)\left(1 - \zeta^2\right) + \hat{r}_{\alpha\beta}(x_1, x_2)$$

$$\mu_{\alpha3} = s_\alpha(x_1, x_2)\zeta$$

$$\mu_{3\beta} = 0$$

$$\mu_{33} = 0$$

Kinematic assumptions

$$u_\alpha = \psi_\alpha(x_1, x_2)\zeta$$

$$u_3 = w^*(x_1, x_2)\left(1 - \zeta^2\right) + \hat{w}(x_1, x_2)$$

$$\phi_\alpha = \omega_\alpha^*(x_1, x_2)\left(1 - \zeta^2\right) + \hat{\omega}_\alpha(x_1, x_2)$$

$$\phi_3 = \omega_3(x_1, x_2)\zeta$$

3.3 COSSERAT PLATE STATIC EQUILIBRIUM EQUATIONS

We substitute the stress assumptions (3.15)–(3.19) and couple stress assumptions (3.25)–(3.29) discussed in the previous section into the three-dimensional Cosserat equilibrium equations (3.13) and (3.14). This results in the following equilibrium system of equations for Cosserat Plate:

$$M_{\alpha\beta,\alpha} - Q_\beta = 0, \tag{3.34}$$

$$Q^*_{\alpha,\alpha} + p^* = 0, \tag{3.35}$$

$$\hat{Q}_{\alpha,\alpha} + \hat{p} = 0, \tag{3.36}$$

$$R^*_{\alpha\beta,\alpha} + \varepsilon_{3\beta\gamma}\left(Q^*_\gamma - Q_\gamma\right) = 0, \tag{3.37}$$

$$\hat{R}_{\alpha\beta,\alpha} + \varepsilon_{3\beta\gamma}\hat{Q}_\gamma = 0, \tag{3.38}$$

$$S_{\alpha,\alpha} + \varepsilon_{3\beta\gamma}M_{\beta\gamma} = 0. \tag{3.39}$$

Here

$$M_{11}, M_{22} - \text{bending moments,}$$
$$M_{12}, M_{21} - \text{twisting moments,}$$
$$Q_\alpha - \text{shear forces,}$$
$$Q^*_\alpha, \hat{Q}_\alpha - \text{transverse shear forces,}$$
$$R^*_{11}, R^*_{22}, \hat{R}_{11}, \hat{R}_{22} - \text{Cosserat bending moments,}$$
$$R^*_{12}, R^*_{21}, \hat{R}_{12}, \hat{R}_{21} - \text{Cosserat twisting moments,}$$
$$S_\alpha - \text{Cosserat couple moments,}$$

are defined the per unit length as follows:

$$M_{\alpha\beta} = \frac{h^2}{12}\int_{-1}^{1} m_{\alpha\beta}d\zeta = \frac{h^2}{6}m_{\alpha\beta},$$

$$Q_\alpha = \frac{h}{2}\int_{-1}^{1} q_\alpha\left(1 - \zeta^2\right)d\zeta = \frac{2h}{3}q_\alpha,$$

$$Q^*_\alpha = \frac{h}{2}\int_{-1}^{1} q^*_\alpha\left(1 - \zeta^2\right)d\zeta = \frac{2h}{3}q^*_\alpha,$$

$$\hat{Q}_\alpha = \frac{h}{2}\int_{-1}^{1} \hat{q}_\alpha\left(1 - \zeta^2\right)d\zeta = \frac{2h}{3}\hat{q}_\alpha,$$

$$R^*_{\alpha\beta} = \frac{h}{2}\int_{-1}^{1} r^*_{\alpha\beta}\left(1 - \zeta^2\right)d\zeta = \frac{2h}{3}r^*_{\alpha\beta},$$

$$\hat{R}_{\alpha\beta} = \frac{h}{2}\int_{-1}^{1} \hat{r}_{\alpha\beta}\left(1 - \zeta^2\right)d\zeta = \frac{2h}{3}\hat{r}_{\alpha\beta},$$

$$S_\alpha = \frac{h^2}{12}\int_{-1}^{1} s_\alpha d\zeta = \frac{h^2}{6}s_\alpha.$$

The pressures p^* and $\hat{p}$ are given as

$$p_1^*(x_1, x_2) = \eta p(x_1, x_2),$$
$$\hat{p}_1(x_1, x_2) = \frac{2}{3}(1 - \eta)p(x_1, x_2).$$

Cosserat Plate Static Equilibrium Equations

$$M_{\alpha\beta,\alpha} - Q_\beta = 0$$
$$Q_{\alpha,\alpha}^* + p^* = 0$$
$$\hat{Q}_{\alpha,\alpha} + \hat{p} = 0$$
$$R_{\alpha\beta,\alpha}^* + \varepsilon_{3\beta\gamma}\left(Q_\gamma^* - Q_\gamma\right) = 0$$
$$\hat{R}_{\alpha\beta,\alpha} + \varepsilon_{3\beta\gamma}\hat{Q}_\gamma = 0$$
$$S_{\alpha,\alpha} + \varepsilon_{3\beta\gamma}M_{\beta\gamma} = 0$$

3.4   STRESS ENERGY OF THE COSSERAT PLATE

We define the plate stress energy by the formula

$$\Phi(\mathcal{S}, \eta) = \frac{h}{2}\int_{-1}^{1} \Phi\{\sigma, \mu\}\, d\zeta, \qquad (3.40)$$

where

$$\Phi\{\sigma, \mu\} = \frac{\mu' + \alpha'}{2}\sigma_{ij}\sigma_{ij} + \frac{\mu' - \alpha'}{2}\sigma_{ij}\sigma_{ji} + \frac{\lambda'}{2}\sigma_{kk}\sigma_{nn}$$
$$+ \frac{\gamma' + \epsilon'}{2}\mu_{ij}\mu_{ij} + \frac{\gamma' - \epsilon'}{2}\mu_{ij}\mu_{ji} + \frac{\beta'}{2}\mu_{kk}\mu_{nn}. \qquad (3.41)$$

Assuming that $\sigma^b(x_1, x_2) = 0$ in the equations (3.23) and (3.24), we have

$$\sigma_0(x_1, x_2) = \frac{1}{2}p(x_1, x_2),$$

Taking into account the stress assumptions (3.15)–(3.19) and couple stress assumptions (3.25)–(3.29), we integrate $\Phi\{\sigma, \mu\}$ with respect to ζ on $[-1, 1]$ and obtain the explicit plate stress energy

$$\Phi(\mathcal{S},\eta) = \frac{3\lambda}{h^3\mu(3\lambda+2\mu)}M_{\alpha\alpha}M_{\beta\beta} + \frac{3(\alpha+\mu)}{2h^3\alpha\mu}M_{\alpha\beta}^2$$

$$+ \frac{3(\alpha-\mu)}{2h^3\alpha\mu}M_{\alpha\beta}M_{\beta\alpha}$$

$$+ \frac{\alpha-\mu}{280h^3\alpha\mu}\left[21Q_\alpha\left(5\hat{Q}_\alpha + 4Q_\alpha^*\right)\right]$$

$$+ \frac{3(\alpha+\mu)}{160h^3\alpha\mu}\left[8\hat{Q}_\alpha\hat{Q}_\alpha + 15Q_\alpha\hat{Q}_\alpha + 20\hat{Q}_\alpha Q_\alpha^* + 8Q_\alpha^*Q_\alpha^*\right]$$

$$- \frac{\gamma-\epsilon}{160h\gamma\epsilon}\left[24R_{\alpha\alpha}^{2*} + 45\hat{R}_{\alpha\alpha} + 60R_{\alpha\beta}^*\hat{R}_{\alpha\beta} + 48R_{12}^*R_{21}^*\right]$$

$$+ \frac{\gamma+\epsilon}{160h^3\gamma\epsilon}\left[8R_{\alpha\beta}^{2*} + 15\hat{R}_{\alpha\beta}\hat{R}_{\alpha\beta} + 20R_{\alpha\beta}^*\hat{R}_{\alpha\beta}\right]$$

$$+ \frac{3\beta}{80h\gamma(3\beta+2\gamma)}\left[8R_{\alpha\alpha}^*R_{\beta\beta}^* + 15\hat{R}_{\alpha\alpha}\hat{R}_{\beta\beta} + 20R_{\alpha\alpha}^*\hat{R}_{\alpha\alpha}\right]$$

$$+ \frac{3(\gamma+\epsilon)}{2h^3\gamma\epsilon}S_\alpha^2 - \frac{\lambda(5p^*+6\hat{p})}{10h\mu(3\lambda+2\mu)}M_{\alpha\alpha}$$

$$+ \frac{(\lambda+\mu)h}{20\mu(3\lambda+2\mu)}\left(2+5\eta-5\eta^2\right)\hat{p}p^*$$

$$+ \frac{(\lambda+\mu)h}{280\mu(3\lambda+2\mu)}\left(17+35\eta^2\right)p^{*2}$$

$$+ \frac{(\lambda+\mu)h}{24\mu(3\lambda+2\mu)}\left(4-6\eta+3\eta^2\right)\hat{p}^2 \tag{3.42}$$

where

$$p^* = \eta p(x_1,x_2,t),$$
$$\hat{p} = (1-\eta)p(x_1,x_2,t),$$

and η is the splitting parameter. We define the Cosserat stress set $\mathcal{S}$ as

$$\mathcal{S} = \left[M_{\alpha\beta}, Q_\alpha, Q_\alpha^*, \hat{Q}_\alpha, R_{\alpha\beta}^*, \hat{R}_{\alpha\beta}, S_\beta\right]. \tag{3.43}$$

The stress energy of the plate P is defined as

$$U_K^{\mathcal{S}} = \int_P \Phi(\mathcal{S},\eta)da. \tag{3.44}$$

3.5 STRAIN-DISPLACEMENT AND TORSION-MICROROTATION RELATIONS

The three-dimensional strain-displacement and torsion-microrotation relations are given as

$$\gamma_{ji} = u_{i,j} + \varepsilon_{ijk}\phi_k, \tag{3.45}$$

$$\chi_{ji} = \phi_{i,j}. \tag{3.46}$$

Let us define the work done by the stress and couple stress over the Cosserat strain field as

$$\mathcal{W} = \frac{h}{2} \int_{-1}^{1} (\sigma \cdot \gamma + \mu \cdot \chi)\, d\zeta. \tag{3.47}$$

We will approximate the plate energy as

$$\mathcal{W} = \mathcal{S} \cdot \mathcal{E} = M_{\alpha\beta} e_{\alpha\beta} + Q_\alpha \pi_\alpha + Q_\alpha^* \pi_\alpha^* + \hat{Q}_\alpha \hat{\pi}_\alpha + R_{\alpha\beta}^* \tau_{\alpha\beta}^* + \hat{R}_{\alpha\beta} \hat{\tau}_{\alpha\beta} + S_\alpha \tau_{3\alpha},$$

where $\mathcal{E}$ is the Cosserat plate strain set

$$\mathcal{E} = \left[e_{\alpha\beta}, \pi_\alpha, \pi_\alpha^*, \hat{\pi}_\alpha, \tau_{\alpha\beta}^*, \hat{\tau}_{\alpha\beta}, \tau_{3\alpha} \right].$$

with the components of $\mathcal{E}$ defined as follows: we substitute the stress and couple stress assumptions into the product $\sigma_{ij}\gamma_{ij}$ and integrate the expression by ζ.

We substitute the Cosserat plate kinematic assumptions (3.30)–(3.33) into the three-dimensional strain-displacement and torsion-microrotation relations (3.45) and (3.46) and obtain the components of Cosserat plate strain in terms of the components of the set of the kinematic variables $\mathcal{U}$ as the following Cosserat plate strain-displacement and torsion-microrotation relations:

$$e_{\alpha\beta} = \Psi_{\beta,\alpha} + \varepsilon_{3\alpha\beta}\Omega_3, \tag{3.48}$$

$$\pi_\alpha = \Psi_\alpha + \varepsilon_{3\alpha\beta}\Omega_\beta^* + \varepsilon_{3\alpha\beta}\hat{\Omega}_\beta, \tag{3.49}$$

$$\pi_\alpha^* = \hat{W}_{,\alpha} + W_{,\alpha}^* + \varepsilon_{3\alpha\beta}\Omega_\beta^* + \varepsilon_{3\alpha\beta}\hat{\Omega}_\beta, \tag{3.50}$$

$$\hat{\pi}_\alpha = k_2 \hat{W}_{,\alpha} + k_1 W_{,\alpha}^* + k_1 \varepsilon_{3\alpha\beta}\Omega_\beta^* + k_2 \varepsilon_{3\alpha\beta}\hat{\Omega}_\beta, \tag{3.51}$$

$$\tau_{\alpha\beta}^* = \Omega_{\beta,\alpha}^* + \hat{\Omega}_{\beta,\alpha}, \tag{3.52}$$

$$\hat{\tau}_{\alpha\beta} = k_1 \Omega_{\beta,\alpha}^* + k_2 \hat{\Omega}_{\beta,\alpha}, \tag{3.53}$$

$$\tau_{3\alpha} = \Omega_{3,\alpha}, \tag{3.54}$$

where $k_1 = \frac{5}{4}$, $k_2 = \frac{3}{2}$.

We have

$$\Psi_\alpha = \frac{1}{h} \int_{-1}^{1} \psi_\alpha(x_1, x_2)\, d\zeta = \frac{2}{h} \psi_\alpha(x_1, x_2),$$

$$W^* = \frac{3}{5} \int_{-1}^{1} w^*(x_1, x_2)\left(1 - \zeta^2\right) d\zeta = \frac{4}{5} w^*(x_1, x_2),$$

$$\hat{W} = \frac{3}{4} \int_{-1}^{1} \hat{w}(x_1, x_2)\left(1 - \zeta^2\right) d\zeta = \hat{w}(x_1, x_2),$$

$$\Omega_\alpha^* = \frac{3}{5} \int_{-1}^{1} \omega_\alpha^*(x_1, x_2)\left(1 - \zeta^2\right) d\zeta = \frac{4}{5} \omega_\alpha^*(x_1, x_2),$$

$$\hat{\Omega}_\alpha = \frac{3}{4} \int_{-1}^{1} \hat{\omega}_\alpha(x_1, x_2)\left(1 - \zeta^2\right) d\zeta = \hat{\omega}_\alpha(x_1, x_2),$$

$$\Omega_3 = \frac{1}{h} \int_{-1}^{1} \omega_3(x_1, x_2)\, d\zeta = \frac{2}{h} \omega_3(x_1, x_2),$$

where Ψ_α are the rotations of the middle plane around x_α axis, $W^* + \hat{W}$ is the vertical deflection of the middle plate, $\hat{\Omega}_\alpha + \Omega_\alpha^*$ are the microrotations in the middle plane around x_α axis, Ω_3 is the rate of change of the microrotation ϕ_3 along x_3.

Therefore, we can introduce the set of Cosserat kinematic variables defined as

$$\mathcal{U} = \left[\Psi_\alpha, W^*, \hat{W}, \Omega_\alpha^*, \hat{\Omega}_\alpha, \Omega_3\right].$$

Cosserat Plate Strain-Displacement and Torsion-Microrotation Relations

$$e_{\alpha\beta} = \Psi_{\beta,\alpha} + \varepsilon_{3\alpha\beta}\Omega_3$$

$$\pi_\alpha = \Psi_\alpha + \varepsilon_{3\alpha\beta}\Omega_\beta^* + \varepsilon_{3\alpha\beta}\hat{\Omega}_\beta$$

$$\pi_\alpha^* = \hat{W}_{,\alpha} + W_{,\alpha}^* + \varepsilon_{3\alpha\beta}\Omega_\beta^* + \varepsilon_{3\alpha\beta}\hat{\Omega}_\beta$$

$$\hat{\pi}_\alpha = k_2 \hat{W}_{,\alpha} + k_1 W_{,\alpha}^* + k_1 \varepsilon_{3\alpha\beta}\Omega_\beta^* + k_2 \varepsilon_{3\alpha\beta}\hat{\Omega}_\beta$$

$$\tau_{\alpha\beta}^* = \Omega_{\beta,\alpha}^* + \hat{\Omega}_{\beta,\alpha}$$

$$\hat{\tau}_{\alpha\beta} = k_1 \Omega_{\beta,\alpha}^* + k_2 \hat{\Omega}_{\beta,\alpha}$$

$$\tau_{3\alpha} = \Omega_{3,\alpha}$$

3.6 CONSTITUTIVE FORMULAS FOR COSSERAT PLATE

We differentiate the plate stress energy (3.42) by the components of the Cosserat plate stress set S and obtain the following constitutive formulas for Cosserat Plate[1]:

$$e_{\alpha\alpha} = \frac{12(\lambda + \mu)}{h^3 \mu (3\lambda + 2\mu)} M_{\alpha\alpha} - |\varepsilon_{\alpha\beta3}| \frac{6\lambda}{h^3 \mu (3\lambda + 2\mu)} M_{\beta\beta}$$
$$- \frac{\lambda (6\hat{p}^* + 5\hat{p})}{10h\mu(3\lambda + 2\mu)}, \tag{3.55}$$

$$e_{\alpha\beta} = \frac{3(\alpha + \mu)}{h^3 \alpha \mu} M_{\alpha\beta} + \frac{3(\alpha - \mu)}{h^3 \alpha \mu} M_{\beta\alpha}, \tag{3.56}$$

$$\pi_\alpha = \frac{3(\alpha - \mu)}{10h\alpha\mu} Q_\alpha^* + \frac{3(\alpha + \mu)}{10h\alpha\mu} Q_\alpha + \frac{3(\alpha - \mu)}{8h\alpha\mu} \hat{Q}_\alpha, \tag{3.57}$$

$$\pi_\alpha^* = \frac{3(\alpha - \mu)}{10h\alpha\mu} Q_\alpha + \frac{3(\alpha + \mu)}{10h\alpha\mu} Q_\alpha^* + \frac{3(\alpha + \mu)}{8h\alpha\mu} \hat{Q}_\alpha, \tag{3.58}$$

$$+ \frac{3(\beta + \gamma)}{2h\gamma(3\beta + 2\gamma)} \hat{R}_{\alpha\alpha} - |\varepsilon_{\alpha\beta3}| \frac{3\beta}{4h\gamma(3\beta + 2\gamma)} R_{\beta\beta}^*, \tag{3.59}$$

$$\hat{\pi}_\alpha = \frac{3(\alpha - \mu)}{8h\alpha\mu} Q_\alpha + \frac{3(\alpha + \mu)}{8h\alpha\mu} Q_\alpha^* + \frac{9(\alpha + \mu)}{16h\alpha\mu} \hat{Q}_\alpha, \tag{3.60}$$

$$\tau_{\alpha\alpha}^* = \frac{6(\beta + \gamma)}{5h\gamma(3\beta + 2\gamma)} R_{\alpha\alpha}^* - |\varepsilon_{\alpha\beta3}| \frac{3\beta}{5h\gamma(3\beta + 2\gamma)} \hat{R}_{\beta\beta},$$

$$\tau_{\alpha\beta}^* = \frac{3(\epsilon + \gamma)}{10h\gamma\epsilon} R_{\alpha\beta}^* - \frac{3(\gamma - \epsilon)}{10h\gamma\epsilon} R_{\beta\alpha}^*$$
$$+ \frac{3(\epsilon + \gamma)}{8h\gamma\epsilon} \hat{R}_{\alpha\beta} - \frac{3(\gamma - \epsilon)}{8h\gamma\epsilon} \hat{R}_{\beta\alpha}, \tag{3.61}$$

$$\hat{\tau}_{\alpha\alpha} = \frac{3(\beta + \gamma)}{2h\gamma(3\beta + 2\gamma)} R_{\alpha\alpha}^* - |\varepsilon_{\alpha\beta3}| \frac{9\beta}{8h\gamma(3\beta + 2\gamma)} \hat{R}_{\beta\beta}$$
$$+ \frac{3(\beta + \gamma)}{4h\gamma(3\beta + 2\gamma)} \hat{R}_{\alpha\alpha} |\varepsilon_{\alpha\beta3}| \frac{3\beta}{4h\gamma(3\beta + 2\gamma)} R_{\beta\beta}^*, \tag{3.62}$$

$$\hat{\tau}_{\alpha\beta} = \frac{3(\epsilon + \gamma)}{8h\gamma\epsilon} R_{\alpha\beta}^* - \frac{3(\gamma - \epsilon)}{8h\gamma\epsilon} R_{\beta\alpha}^*$$
$$+ \frac{3(\epsilon + \gamma)}{16h\gamma\epsilon} \hat{R}_{\alpha\beta} - \frac{3(\gamma - \epsilon)}{16h\gamma\epsilon} \hat{R}_{\beta\alpha}, \tag{3.63}$$

$$\tau_{3\alpha} = \frac{3(\gamma + \epsilon)}{h^3 \gamma\epsilon} S_\alpha. \tag{3.64}$$

---

[1]In the following formulas a subindex $\beta = 1$ iff $\alpha = 2$ and $\beta = 2$ iff $\alpha = 1$.

Cosserat Plate Constitutive Formulas

$$e_{\alpha\alpha} = \frac{12(\lambda + \mu)}{h^3 \mu (3\lambda + 2\mu)} M_{\alpha\alpha} - |\varepsilon_{\alpha\beta3}| \frac{6\lambda}{h^3 \mu (3\lambda + 2\mu)} M_{\beta\beta}$$
$$- \frac{\lambda(6p^* + 5\hat{p})}{10h\mu(3\lambda + 2\mu)}$$

$$e_{\alpha\beta} = \frac{3(\alpha + \mu)}{h^3 \alpha \mu} M_{\alpha\beta} + \frac{3(\alpha - \mu)}{h^3 \alpha \mu} M_{\beta\alpha}$$

$$\pi_\alpha = \frac{3(\alpha - \mu)}{10h\alpha\mu} Q_\alpha^* + \frac{3(\alpha + \mu)}{10h\alpha\mu} Q_\alpha + \frac{3(\alpha - \mu)}{8h\alpha\mu} \hat{Q}_\alpha$$

$$\pi_\alpha^* = \frac{3(\alpha - \mu)}{10h\alpha\mu} Q_\alpha + \frac{3(\alpha + \mu)}{10h\alpha\mu} Q_\alpha^* + \frac{3(\alpha + \mu)}{8h\alpha\mu} \hat{Q}_\alpha$$
$$+ \frac{3(\beta + \gamma)}{2h\gamma(3\beta + 2\gamma)} \hat{R}_{\alpha\alpha} - |\varepsilon_{\alpha\beta3}| \frac{3\beta}{4h\gamma(3\beta + 2\gamma)} R_{\beta\beta}^*$$

$$\hat{\pi}_\alpha = \frac{3(\alpha - \mu)}{8h\alpha\mu} Q_\alpha + \frac{3(\alpha + \mu)}{8h\alpha\mu} Q_\alpha^* + \frac{9(\alpha + \mu)}{16h\alpha\mu} \hat{Q}_\alpha$$

$$\tau_{\alpha\alpha}^* = \frac{6(\beta + \gamma)}{5h\gamma(3\beta + 2\gamma)} R_{\alpha\alpha}^* - |\varepsilon_{\alpha\beta3}| \frac{3\beta}{5h\gamma(3\beta + 2\gamma)} \hat{R}_{\beta\beta}$$

$$\tau_{\alpha\beta}^* = \frac{3(\epsilon + \gamma)}{10h\gamma\epsilon} R_{\alpha\beta}^* - \frac{3(\gamma - \epsilon)}{10h\gamma\epsilon} R_{\beta\alpha}^*$$
$$+ \frac{3(\epsilon + \gamma)}{8h\gamma\epsilon} \hat{R}_{\alpha\beta} - \frac{3(\gamma - \epsilon)}{8h\gamma\epsilon} \hat{R}_{\beta\alpha}$$

$$\hat{\tau}_{\alpha\alpha} = \frac{3(\beta + \gamma)}{2h\gamma(3\beta + 2\gamma)} R_{\alpha\alpha}^* - |\varepsilon_{\alpha\beta3}| \frac{9\beta}{8h\gamma(3\beta + 2\gamma)} \hat{R}_{\beta\beta}$$
$$+ \frac{3(\beta + \gamma)}{4h\gamma(3\beta + 2\gamma)} \hat{R}_{\alpha\alpha} |\varepsilon_{\alpha\beta3}| \frac{3\beta}{4h\gamma(3\beta + 2\gamma)} R_{\beta\beta}^*$$

$$\hat{\tau}_{\alpha\beta} = \frac{3(\epsilon + \gamma)}{8h\gamma\epsilon} R_{\alpha\beta}^* - \frac{3(\gamma - \epsilon)}{8h\gamma\epsilon} R_{\beta\alpha}^*$$
$$+ \frac{3(\epsilon + \gamma)}{16h\gamma\epsilon} \hat{R}_{\alpha\beta} - \frac{3(\gamma - \epsilon)}{16h\gamma\epsilon} \hat{R}_{\beta\alpha}$$

$$\tau_{3\alpha} = \frac{3(\gamma + \epsilon)}{h^3 \gamma\epsilon} S_\alpha$$

3.7 VARIATIONAL FORMULATION OF COSSERAT PLATE BENDING

We are looking for the components of the stress set $\mathcal{S}$, the strain set $\mathcal{E}$ and the set of kinematic variables $\mathcal{U}$ satisfy the strain-displacement and torsion-microrotation relations (3.48)–(3.54), the constitutive formulas (3.55)–(3.64) and the equilibrium system of equations (3.34)–(3.39) with the traction boundary conditions[2]

$$M_{\alpha\beta}n_\beta = M_{o\alpha} \text{ on } \Gamma_\sigma, \tag{3.65}$$

$$Q_\alpha^* n_\alpha = Q_o^* \text{ on } \Gamma_\sigma, \tag{3.66}$$

$$\hat{Q}_\alpha n_\alpha = \hat{Q}_o \text{ on } \Gamma_\sigma, \tag{3.67}$$

$$R_{\alpha\beta}^* n_\beta = R_{o\alpha}^* \text{ on } \Gamma_\sigma, \tag{3.68}$$

$$\hat{R}_{\alpha\beta}n_\beta = \hat{R}_{o\alpha} \text{ on } \Gamma_\sigma, \tag{3.69}$$

$$S_\alpha n_\alpha = S_o \text{ on } \Gamma_\sigma, \tag{3.70}$$

and the displacement boundary conditions

$$\Psi_\alpha = \Psi_{o\alpha} \text{ on } \Gamma_u, \tag{3.71}$$

$$W^* = W_o^* \text{ on } \Gamma_u, \tag{3.72}$$

$$\hat{W} = \hat{W}_o \text{ on } \Gamma_u, \tag{3.73}$$

$$\Omega_\alpha^* = \Omega_{o\alpha}^* \text{ on } \Gamma_u, \tag{3.74}$$

$$\hat{\Omega}_\alpha = \hat{\Omega}_{o\alpha} \text{ on } \Gamma_u, \tag{3.75}$$

$$\Omega_3 = \Omega_{o3} \text{ on } \Gamma_u, \tag{3.76}$$

and minimize the bending plate stress energy $U_K^{\mathcal{S}}$

$$U_K^{\mathcal{S}} = \int_{P_0} \Phi(\mathcal{S}, \eta) da.$$

with respect to the splitting parameter η.

3.8 OPTIMAL VALUE OF THE SPLITTING PARAMETER

The equilibrium systems of partial differential equations correspond to a state of the system (3.93), where the minimum of the

---

[2]The vector n_β is the outward unit normal vector to Γ_σ.

energy is reached. The optimization of the splitting parameter appears as a result of the variational principle for the Cosserat elastic plate (3.7). The bending system of equations depends on the splitting parameter and therefore its solution is parametric. The following minimization procedure for the elastic energy allows us to find the optimal value of this parameter, which corresponds to the unique solution of the bending problem [42].

Let $\mathcal{W}^{(\eta)}$ be the work done by the stress and couple stress over the strain field. The optimal value of the splitting parameter η is the minimizer of the work $\mathcal{W}^{(\eta)}$. Let us now obtain the explicit expression for the optimal value of the parameter η.

Let $\mathcal{S}^{(0)}$ and $\mathcal{E}^{(0)}$ be the stress and strain sets that correspond to the value of $\eta = 0$ and $\mathcal{S}^{(1)}$ and $\mathcal{E}^{(1)}$ be the stress and strain sets that correspond to the value of $\eta = 1$. Let us represent each component of the stress set $\mathcal{S}^{(\eta)}$ and the strain set $\mathcal{E}^{(\eta)}$ as a linear combinations of the stress sets $\mathcal{S}^{(0)}$ and $\mathcal{S}^{(1)}$, and the strain sets $\mathcal{E}^{(0)}$ and $\mathcal{E}^{(1)}$ respectively:

$$\mathcal{S}^{(\eta)} = (1-\eta)\mathcal{S}^{(0)} + \eta\mathcal{S}^{(1)},$$
$$\mathcal{E}^{(\eta)} = (1-\eta)\mathcal{E}^{(0)} + \eta\mathcal{E}^{(1)}.$$

Therefore, the work $\mathcal{W}^{(\eta)}$ done by the stress and couple stress over the strain field

$$\mathcal{W}^{(\eta)} = \mathcal{S}^{(\eta)} \cdot \mathcal{E}^{(\eta)},$$

can be written as

$$\mathcal{W}^{(\eta)} = \left((1-\eta)\mathcal{S}^{(0)} + \eta\mathcal{S}^{(1)}\right) \cdot \left((1-\eta)\mathcal{E}^{(0)} + \eta\mathcal{E}^{(1)}\right).$$

The derivative of the work $\mathcal{W}^{(\eta)}$ by the parameter η can be found as follows:

$$\frac{\partial \mathcal{W}^{(\eta)}}{\partial \eta} = \left(-\mathcal{S}^{(0)} + \mathcal{S}^{(1)}\right) \cdot \left((1-\eta)\mathcal{E}^{(0)} + \eta\mathcal{E}^{(1)}\right) +$$
$$\left((1-\eta)\mathcal{S}^{(0)} + \eta\mathcal{S}^{(1)}\right) \cdot \left(\mathcal{E}^{(1)} - \mathcal{E}^{(0)}\right).$$

The zero of the derivative $\frac{\partial \mathcal{W}^{(\eta)}}{\partial \eta}$ gives the optimal value η_{opt} of the splitting parameter η [42]:

$$\eta_{\text{opt}} = \frac{2\mathcal{W}^{(00)} - \mathcal{W}^{(10)} - \mathcal{W}^{(01)}}{2\left(\mathcal{W}^{(11)} + \mathcal{W}^{(00)} - \mathcal{W}^{(10)} - \mathcal{W}^{(01)}\right)}, \tag{3.77}$$

where

$$\mathcal{W}^{(00)} = \mathcal{S}^{(0)} \cdot \mathcal{E}^{(0)}, \tag{3.78}$$

$$\mathcal{W}^{(01)} = \mathcal{S}^{(0)} \cdot \mathcal{E}^{(1)}, \tag{3.79}$$

$$\mathcal{W}^{(10)} = \mathcal{S}^{(1)} \cdot \mathcal{E}^{(0)}, \tag{3.80}$$

$$\mathcal{W}^{(11)} = \mathcal{S}^{(1)} \cdot \mathcal{E}^{(1)}, \tag{3.81}$$

3.9 STATICS FIELD EQUATIONS

The constitutive formulas for the Cosserat Plate can also be written in the following reverse form by solving the equations (3.55)–(3.64) for the components of the plate stresses and then substituting the expressions for the components of strain and torsion tensors from the equations (3.48)–(3.54)[3]:

$$M_{\alpha\alpha} = \frac{h^3 \mu (\lambda + \mu)}{3(\lambda + 2\mu)} \Psi_{\alpha,\alpha} + \frac{\lambda \mu h^3}{6(\lambda + 2\mu)} \Psi_{\beta,\beta}$$
$$+ \frac{(6p^* + 5\hat{p})\lambda h^2}{60(\lambda + 2\mu)}, \tag{3.82}$$

$$M_{\beta\alpha} = \frac{(\mu - \alpha)h^3}{12} \Psi_{\alpha,\beta} + \frac{h^3(\alpha + \mu)}{12} \Psi_{\beta,\alpha} +$$
$$(-1)^\beta \frac{\alpha h^3}{6} \Omega_3, \tag{3.83}$$

$$Q_\alpha = \frac{5h(\alpha + \mu)}{6} \Psi_\alpha + \frac{5(\mu - \alpha)h}{6} \hat{W}_{,\alpha}$$
$$+ \frac{5(\mu - \alpha)h}{6} W^*_{,\alpha} + (-1)^\beta \frac{5h\alpha}{3} \Omega^*_\beta \tag{3.84}$$
$$+ (-1)^\beta \frac{5h\alpha}{3} \hat{\Omega}_\beta, \tag{3.85}$$

$$Q^*_\alpha = \frac{5(\mu - \alpha)h}{6} \Psi_\alpha + \frac{5(\mu - \alpha)^2 h}{6(\mu + \alpha)} \hat{W}_{,\alpha}$$
$$+ \frac{5(\mu + \alpha)h}{6} W^*_{,\alpha} + (-1)^\alpha \frac{5h\alpha}{3} \Omega^*_\beta$$
$$+ (-1)^\alpha \frac{5h\alpha(\mu - \alpha)}{3(\mu + \alpha)} \hat{\Omega}_\beta, \tag{3.86}$$

---

[3]In the following formulas a subindex $\beta = 1$ iff $\alpha = 2$ and $\beta = 2$ iff $\alpha = 1$.

$$\hat{Q}_\alpha = \frac{8\alpha\mu h}{3(\mu+\alpha)}\hat{W}_{,\alpha} + (-1)^\alpha \frac{8\alpha\mu h}{3(\mu+\alpha)}\hat{\Omega}_\beta, \tag{3.87}$$

$$R^*_{\alpha\alpha} = \frac{10h\gamma(\beta+\gamma)}{3(\beta+2\gamma)}\Omega^*_{\alpha,\alpha} + \frac{5h\beta\gamma}{3(\beta+2\gamma)}\Omega^*_{\beta,\beta}, \tag{3.88}$$

$$R^*_{\beta\alpha} = \frac{5(\gamma-\epsilon)h}{6}\Omega^*_{\beta,\alpha} + \frac{5h(\gamma+\epsilon)}{6}\Omega^*_{\alpha,\beta}, \tag{3.89}$$

$$\hat{R}_{\alpha\alpha} = \frac{8\gamma(\gamma+\beta)h}{3(\beta+2\gamma)}\hat{\Omega}_{\alpha,\alpha} + \frac{4\gamma\beta h}{3(\beta+2\gamma)}\hat{\Omega}_{\beta,\beta}, \tag{3.90}$$

$$\hat{R}_{\beta\alpha} = \frac{2(\gamma-\epsilon)h}{3}\hat{\Omega}_{\beta,\alpha} + \frac{2(\gamma+\epsilon)h}{3}\hat{\Omega}_{\alpha,\beta}, \tag{3.91}$$

$$S_\alpha = \frac{\gamma\epsilon h^3}{3(\gamma+\epsilon)}\Omega_{3,\alpha}. \tag{3.92}$$

In order to obtain the Cosserat plate bending field equations in terms of the kinematic variables, we substitute the constitutive formulas in the reverse form (3.82)–(3.92) into the bending system of equations (3.34)–(3.39). If the solution vector $\mathcal{U}$ of the kinematic variables be defined as

$$\mathcal{U} = \left[\Psi_1, \Psi_2, W, \Omega_3, \Omega_1, \Omega_2, W^*, \hat{\Omega}_1, \hat{\Omega}_2\right]^T,$$

then the Cosserat plate bending field equations can be written in the following form

$$L\mathcal{U} = f(\eta). \tag{3.93}$$

The operator L here is given as

$$\begin{bmatrix}
L_{11} & L_{12} & L_{13} & L_{14} & 0 & L_{16} & 0 & L_{18} & L_{19} \\
L_{12} & L_{22} & L_{23} & L_{24} & L_{25} & 0 & L_{27} & 0 & L_{29} \\
L_{31} & L_{32} & L_{33} & L_{34} & L_{35} & L_{36} & L_{37} & L_{38} & 0 \\
L_{41} & L_{42} & L_{43} & L_{44} & L_{45} & L_{46} & L_{47} & L_{48} & 0 \\
0 & L_{52} & L_{53} & L_{54} & L_{55} & L_{56} & L_{57} & 0 & 0 \\
L_{61} & 0 & L_{63} & L_{64} & L_{65} & L_{66} & 0 & L_{68} & 0 \\
0 & 0 & 0 & L_{74} & 0 & 0 & L_{77} & L_{78} & 0 \\
0 & 0 & 0 & L_{84} & 0 & 0 & L_{87} & L_{88} & 0 \\
L_{91} & L_{92} & 0 & 0 & 0 & 0 & 0 & 0 & L_{99}
\end{bmatrix}$$

and the right-hand side $f(\eta)$ vector is

$$f(\eta) = \begin{bmatrix} -\frac{h^3\lambda(6p^*_{,1}+5\hat{p}_{,1})}{120(\lambda+2\mu)} \\ -\frac{h^3\lambda(6p^*_{,2}+5\hat{p}_{,2})}{120(\lambda+2\mu)} \\ -\frac{6p^*+5\hat{p}}{6} \\ -p^* \\ 0 \\ 0 \\ 0 \\ 0 \\ 0 \end{bmatrix},$$

where the pressures p^* and $\hat{p}$ are defined as before:

$$p^*(x_1,x_2) = \eta\, p(x_1,x_2), \tag{3.94}$$

$$\hat{p}(x_1,x_2) = (1-\eta)\, p(x_1,x_2). \tag{3.95}$$

The operators L_{ij} are defined as follows:

$$L_{11} = c_1\frac{\partial^2}{\partial x_1^2} + c_2\frac{\partial^2}{\partial x_2^2} - c_3, \qquad L_{12} = c_4\frac{\partial^2}{\partial x_1 x_2},$$

$$L_{13} = c_5\frac{\partial}{\partial x_1}, \qquad L_{14} = c_5\frac{\partial}{\partial x_1},$$

$$L_{16} = c_6, \qquad L_{18} = c_6,$$

$$L_{19} = c_7\frac{\partial}{\partial x_2}, \qquad L_{21} = c_4\frac{\partial^2}{\partial x_1 x_2},$$

$$L_{22} = c_2\frac{\partial^2}{\partial x_1^2} + c_1\frac{\partial^2}{\partial x_2^2} - c_3, \qquad L_{23} = c_5\frac{\partial}{\partial x_2},$$

$$L_{24} = c_5\frac{\partial}{\partial x_2}, \qquad L_{25} = -c_6,$$

$$L_{27} = -c_6, \qquad L_{29} = -c_7\frac{\partial}{\partial x_1},$$

$$L_{31} = c_5\frac{\partial}{\partial x_1}, \qquad L_{32} = -c_5\frac{\partial}{\partial x_1},$$

$$L_{33} = c_3\frac{\partial^2}{\partial x_1^2} + c_3\frac{\partial^2}{\partial x_2^2}, \qquad L_{34} = c_3\frac{\partial^2}{\partial x_1^2} + c_3\frac{\partial^2}{\partial x_2^2},$$

$$L_{35} = -c_6 \frac{\partial}{\partial x_2},$$

$$L_{36} = c_6 \frac{\partial}{\partial x_1},$$

$$L_{37} = -c_6 \frac{\partial}{\partial x_2},$$

$$L_{38} = c_6 \frac{\partial}{\partial x_1},$$

$$L_{41} = c_5 \frac{\partial}{\partial x_1},$$

$$L_{42} = -c_5 \frac{\partial}{\partial x_1},$$

$$L_{43} = c_3 \frac{\partial^2}{\partial x_1^2} + c_3 \frac{\partial^2}{\partial x_2^2},$$

$$L_{44} = c_8 \frac{\partial^2}{\partial x_1^2} + c_8 \frac{\partial^2}{\partial x_2^2},$$

$$L_{45} = -c_6 \frac{\partial}{\partial x_2},$$

$$L_{46} = c_6 \frac{\partial}{\partial x_1},$$

$$L_{47} = -c_9 \frac{\partial}{\partial x_2},$$

$$L_{48} = c_9 \frac{\partial}{\partial x_1},$$

$$L_{52} = -c_6,$$

$$L_{53} = c_6 \frac{\partial}{\partial x_2},$$

$$L_{54} = c_9 \frac{\partial}{\partial x_2},$$

$$L_{55} = c_{10} \frac{\partial^2}{\partial x_1^2} + c_{11} \frac{\partial^2}{\partial x_2^2} - 2c_6,$$

$$L_{56} = c_{12} \frac{\partial^2}{\partial x_1 x_2},$$

$$L_{57} = -c_{13},$$

$$L_{61} = c_6,$$

$$L_{63} = -c_6 \frac{\partial}{\partial x_2},$$

$$L_{64} = -c_9 \frac{\partial}{\partial x_2},$$

$$L_{65} = c_{12} \frac{\partial^2}{\partial x_1 x_2},$$

$$L_{66} = c_{10} \frac{\partial^2}{\partial x_1^2} + c_{11} \frac{\partial^2}{\partial x_2^2} - 2c_6,$$

$$L_{67} = -c_{13},$$

$$L_{74} = c_{14} \frac{\partial}{\partial x_2},$$

$$L_{77} = c_{10} \frac{\partial^2}{\partial x_1^2} + c_{11} \frac{\partial^2}{\partial x_2^2} - c_{14},$$

$$L_{78} = c_{12} \frac{\partial^2}{\partial x_1 x_2},$$

$$L_{84} = c_{14} \frac{\partial}{\partial x_2},$$

$$L_{78} = c_{12} \frac{\partial^2}{\partial x_1 x_2},$$

$$L_{88} = c_{11} \frac{\partial^2}{\partial x_1^2} + c_{10} \frac{\partial^2}{\partial x_2^2} - c_{14},$$

$$L_{91} = -c_7 \frac{\partial}{\partial x_2},$$

$$L_{92} = c_7 \frac{\partial}{\partial x_1},$$

$$L_{99} = c_{15} \frac{\partial^2}{\partial x_1^2} + c_{15} \frac{\partial^2}{\partial x_2^2} - 2c_7,$$

where c_i are the constants given as

$$c_1 = \frac{h^4 \mu(\lambda + \mu)}{6(\lambda + 2\mu)}, \qquad c_2 = \frac{h^4(\alpha + \mu)}{24},$$

$$c_3 = \frac{5h(\alpha + \mu)}{6}, \qquad c_4 = \frac{h^4 (\mu(3\lambda + 2\mu) - \alpha(\lambda + 2\mu))}{24(\lambda + 2\mu)},$$

$$c_5 = \frac{5h(\alpha - \mu)}{6}, \qquad c_6 = \frac{5h\alpha}{3},$$

$$c_7 = \frac{h^4 \alpha}{12}, \qquad c_8 = \frac{5h(\alpha - \mu)^2}{6(\alpha + \mu)},$$

$$c_9 = \frac{5h\alpha(\alpha - \mu)}{3(\alpha + \mu)}, \qquad c_{10} = \frac{10h\gamma(\beta + \gamma)}{3(\beta + 2\gamma)},$$

$$c_{11} = \frac{5h(\gamma + \epsilon)}{6}, \qquad c_{12} = \frac{5h(2\gamma(\gamma - \epsilon) + \beta(3\gamma - \epsilon))}{6(\beta + 2\gamma)},$$

$$c_{13} = \frac{10h\alpha^2}{3(\alpha + \mu)}, \qquad c_{14} = \frac{10h\alpha\mu}{3(\alpha + \mu)},$$

$$c_{15} = \frac{h^4 \gamma\epsilon}{6(\gamma + \epsilon)}.$$

The parametric system (3.93) is an elliptic system of nine partial differential equations, where L is a linear differential operator acting on the set of kinematic variables $\mathcal{U}$.

CHAPTER 4

Cosserat Plate Dynamics

4.1 SCOPE OF THE CHAPTER

This chapter introduces the dynamic Theory of Cosserat Plates. We assume the distributions of the Cosserat stress, couple stress, displacement, and microrotation consistent with the three-dimensional Cosserat Elasticity. We obtain the Cosserat plate equilibrium equations, constitutive formulas, and the parametric system of equilibrium equations, where the optimal value of the parameter minimizes the Cosserat elastic energy. We also present the elliptic system of Cosserat plate field equations written in terms of the kinematic variables.

4.2 PLATE ASSUMPTIONS

First we remind here that elastic motion of a Cosserat body is described by the following local balance laws chapter 1:

$$\sigma_{ji,j} = \frac{\partial p_i}{\partial t}, \tag{4.1}$$

$$\mu_{ji,j} + \varepsilon_{ijk}\sigma_{jk} = \frac{\partial q_i}{\partial t}, \tag{4.2}$$

DOI: 10.1201/9781003190264-4

Here

$$p_i = \rho \frac{\partial u_i}{\partial t}, \qquad (4.3)$$

$$q_i = J_{ji} \frac{\partial \phi_j}{\partial t}, \qquad (4.4)$$

and ρ is the material density, J_{ji} is the microinertia, ε_{ijk} is the Levi-Civita tensor.

4.2.1 Boundary and Initial Conditions

Let B_0 be the Cosserat body and ∂B be its boundary. We split the body boundary ∂B into $\mathcal{G}_u$ and $\mathcal{G}_\sigma$, such that $\mathcal{G}_u \cup \mathcal{G}_\sigma = \partial B$. We prescribe the displacements and microrotations on $\mathcal{G}_u$ and the stress and couple stress on $\mathcal{G}_\sigma$.

The three dimensional Cosserat Linear Theory is based the dynamic equations (4.1) and (4.2), the strain-displacement and torsion-microrotation relations (1.10) and (1.11) and the constitutive equations (1.32) and (1.33) are considered together with the following boundary conditions prescribed on ∂B:

$$u_i = \hat{u}_i \text{ on } \mathcal{G}_u \times [t_0, t], \qquad (4.5)$$

$$\phi_i = \hat{\phi}_i \text{ on } \mathcal{G}_u \times [t_0, t], \qquad (4.6)$$

$$\sigma_{ji} n_i = \hat{\sigma}_j \text{ on } \mathcal{G}_\sigma \times [t_0, t], \qquad (4.7)$$

$$\mu_{ji} n_i = \hat{\mu}_j \text{ on } \mathcal{G}_\sigma \times [t_0, t], \qquad (4.8)$$

and the initial conditions on B_0

$$u_i(x, 0) = u_i^0, \qquad (4.9)$$

$$\phi_i(x, 0) = \phi_i^0, \qquad (4.10)$$

$$\dot{u}_i(x, 0) = \dot{u}_i^0, \qquad (4.11)$$

$$\dot{\phi}_i(x, 0) = \dot{\phi}_i^0, \qquad (4.12)$$

where n_i denotes the outward unit normal vector to ∂B.

Here we consider the Cosserat body B as the plate P, where h is the thickness of the plate and $x_3 = 0$ represent its middle plane. The sets T and B are the top and bottom surfaces contained in the planes $x_3 = h/2$, $x_3 = -h/2$ respectively and the curve Γ is the boundary of the middle plane of the plate.

The set of points $P = \left(\Gamma \times [-\frac{h}{2}, \frac{h}{2}] \right) \cup T \cup B$ forms the entire surface of the plate and $\Gamma_u \times [-\frac{h}{2}, \frac{h}{2}]$ is the lateral part of the boundary where displacements and microrotations are prescribed. The notation $\Gamma_\sigma = \Gamma \backslash \Gamma_u$ of the remainder we use to describe the lateral part of the boundary edge $\Gamma_\sigma \times [-\frac{h}{2}, \frac{h}{2}]$ where stress and couple stress are prescribed. We also use notation P_0 for the middle plane internal domain of the plate.

4.2.2 Stresses and Kinematic Approximations

In this case we consider the time-dependent vertical load and pure twisting momentum boundary conditions at the top and bottom of the plate, which can be written in the form:

$$\sigma_{33}(x_1, x_2, h/2, t) = \sigma^t(x_1, x_2, t), \tag{4.13}$$

$$\sigma_{33}(x_1, x_2, -h/2, t) = \sigma^b(x_1, x_2, t), \tag{4.14}$$

$$\sigma_{3\beta}(x_1, x_2, h/2, t) = 0, \tag{4.15}$$

$$\sigma_{3\beta}(x_1, x_2, -h/2, t) = 0, \tag{4.16}$$

$$\mu_{33}(x_1, x_2, h/2, t) = 0, \tag{4.17}$$

$$\mu_{33}(x_1, x_2, -h/2, t) = 0, \tag{4.18}$$

$$\mu_{3\beta}(x_1, x_2, h/2, t) = 0, \tag{4.19}$$

$$\mu_{3\beta}(x_1, x_2, -h/2, t) = 0, \tag{4.20}$$

where $(x_1, x_2) \in P$.

We will also consider the microinertia tensor $\mathbf{J}$ in the following form:

$$\mathbf{J} = \begin{pmatrix} J_{11} & J_{12} & 0 \\ J_{21} & J_{22} & 0 \\ 0 & 0 & J_{33} \end{pmatrix} \tag{4.21}$$

As it is usual in the dynamics, we approximate the three-dimensional stress tensor σ_{ji} and couple stress tensor μ_{ji} by the time-dependent polynomial expressions, similar to the statics:

$$\sigma_{\alpha\beta} = m_{\alpha\beta}(x_1, x_2, t)\zeta, \tag{4.22}$$

$$\sigma_{3\beta} = q_\beta(x_1, x_2, t)\left(1 - \zeta^2\right), \tag{4.23}$$

$$\sigma_{\alpha 3} = q_\alpha^*(x_1, x_2, t)\left(1 - \zeta^2\right) + \hat{q}_\alpha(x_1, x_2, t), \tag{4.24}$$

$$\sigma_{33} = \frac{3}{4}\left(\zeta - \frac{\zeta^3}{3}\right)p^*(x_1, x_2, t) + \frac{1}{2}\hat{p}(x_1, x_2, t)\zeta + \sigma_0(x_1, x_2, t), \tag{4.25}$$

$$\mu_{\alpha\beta} = r_{\alpha\beta}^*(x_1, x_2, t)\left(1 - \zeta^2\right) + \hat{r}_{\alpha\beta}(x_1, x_2, t), \tag{4.26}$$

$$\mu_{\alpha 3} = s_\alpha(x_1, x_2, t)\zeta, \tag{4.27}$$

$$\mu_{3\beta} = 0, \tag{4.28}$$

$$\mu_{33} = 0, \tag{4.29}$$

where

$$p(x_1, x_2, t) = \sigma^t(x_1, x_2, t) - \sigma^b(x_1, x_2, t), \tag{4.30}$$

$$\sigma_0(x_1, x_2, t) = \frac{1}{2}\left(\sigma^t(x_1, x_2, t) + \sigma^b(x_1, x_2, t)\right). \tag{4.31}$$

The pressures p^* and $\hat{p}$ are chosen in the form:

$$p^*(x_1, x_2, t) = \eta\, p(x_1, x_2, t), \tag{4.32}$$

$$\hat{p}(x_1, x_2, t) = (1 - \eta)\, p(x_1, x_2, t), \tag{4.33}$$

and $\eta \in \mathbb{R}$ is called the **splitting parameter**.

We approximate the three-dimensional displacement vector u_i and microrotation ϕ_i by the time-dependent expressions, similar to the statics:

$$u_\alpha = \psi_\alpha(x_1, x_2, t)\zeta, \tag{4.34}$$

$$u_3 = w^*(x_1, x_2, t)\left(1 - \zeta^2\right) + \hat{w}(x_1, x_2, t), \tag{4.35}$$

$$\phi_\alpha = \omega_\alpha^*(x_1, x_2, t)\left(1 - \zeta^2\right) + \hat{\omega}_\alpha(x_1, x_2, t), \tag{4.36}$$

$$\phi_3 = \omega_3(x_1, x_2, t)\zeta. \tag{4.37}$$

Summary of the Cosserat Plate Assumptions

Stress assumptions

$$\sigma_{\alpha\beta} = m_{\alpha\beta}(x_1, x_2, t)\zeta$$

$$\sigma_{3\beta} = q_\beta(x_1, x_2, t)\left(1 - \zeta^2\right)$$

$$\sigma_{\alpha3} = q_\alpha^*(x_1, x_2, t)\left(1 - \zeta^2\right) + \hat{q}_\alpha(x_1, x_2, t)$$

$$\sigma_{33} = \frac{3}{4}\left(\zeta - \frac{\zeta^3}{3}\right)p^*(x_1, x_2, t) + \frac{1}{2}\hat{p}(x_1, x_2, t)\zeta + \sigma_0(x_1, x_2, t)$$

Couple stress assumptions

$$\mu_{\alpha\beta} = r_{\alpha\beta}^*(x_1, x_2, t)\left(1 - \zeta^2\right) + \hat{r}_{\alpha\beta}(x_1, x_2, t)$$

$$\mu_{\alpha3} = s_\alpha(x_1, x_2, t)\zeta$$

$$\mu_{3\beta} = 0$$

$$\mu_{33} = 0$$

Kinematic assumptions

$$u_\alpha = \psi_\alpha(x_1, x_2, t)\zeta$$

$$u_3 = w^*(x_1, x_2, t)\left(1 - \zeta^2\right) + \hat{w}(x_1, x_2, t)$$

$$\phi_\alpha = \omega_\alpha^*(x_1, x_2, t)\left(1 - \zeta^2\right) + \hat{\omega}_\alpha(x_1, x_2, t)$$

$$\phi_3 = \omega_3(x_1, x_2, t)\zeta$$

We substitute the Cosserat plate kinematic assumptions (4.34)–(4.37) into the three-dimensional strain-displacement and torsion-microrotation relations (1.10) and (1.11) and obtain the components of Cosserat plate strain in terms of the components of the set of the kinematic variables as the following Cosserat plate strain-displacement and torsion-microrotation relations:

$$e_{\alpha\beta} = \Psi_{\beta,\alpha} + \varepsilon_{3\alpha\beta}\Omega_3, \tag{4.38}$$

$$\pi_\alpha = \Psi_\alpha + \varepsilon_{3\alpha\beta}\Omega_\beta^* + \varepsilon_{3\alpha\beta}\hat{\Omega}_\beta, \tag{4.39}$$

$$\pi_\alpha^* = \hat{W}_{,\alpha} + W_{,\alpha}^* + \varepsilon_{3\alpha\beta}\Omega_\beta^* + \varepsilon_{3\alpha\beta}\hat{\Omega}_\beta, \tag{4.40}$$

$$\hat{\pi}_\alpha = k_2\hat{W}_{,\alpha} + k_1 W_{,\alpha}^* + k_1\varepsilon_{3\alpha\beta}\Omega_\beta^* + k_2\varepsilon_{3\alpha\beta}\hat{\Omega}_\beta, \tag{4.41}$$

$$\tau_{\alpha\beta}^* = \Omega_{\beta,\alpha}^* + \hat{\Omega}_{\beta,\alpha}, \tag{4.42}$$

$$\hat{\tau}_{\alpha\beta} = k_1\Omega_{\beta,\alpha}^* + k_2\hat{\Omega}_{\beta,\alpha}, \tag{4.43}$$

$$\tau_{3\alpha} = \Omega_{3,\alpha}, \tag{4.44}$$

where $k_1 = \frac{5}{4}$, $k_2 = \frac{3}{2}$ and

$$\Psi_\alpha = \frac{1}{h}\int_{-1}^1 \psi_\alpha(x_1,x_2,t)\,d\zeta = \frac{2}{h}\psi_\alpha(x_1,x_2,t),$$

$$W^* = \frac{3}{5}\int_{-1}^1 w^*(x_1,x_2,t)\left(1-\zeta^2\right)d\zeta = \frac{4}{5}w^*(x_1,x_2,t),$$

$$\hat{W} = \frac{3}{4}\int_{-1}^1 \hat{w}(x_1,x_2,t)\left(1-\zeta^2\right)d\zeta = \hat{w}(x_1,x_2,t),$$

$$\Omega_\alpha^* = \frac{3}{5}\int_{-1}^1 \omega_\alpha^*(x_1,x_2,t)\left(1-\zeta^2\right)d\zeta = \frac{4}{5}\omega_\alpha^*(x_1,x_2,t),$$

$$\hat{\Omega}_\alpha = \frac{3}{4}\int_{-1}^1 \hat{\omega}_\alpha(x_1,x_2,t)\left(1-\zeta^2\right)d\zeta = \hat{\omega}_\alpha(x_1,x_2,t),$$

$$\Omega_3 = \frac{1}{h}\int_{-1}^1 \omega_3(x_1,x_2,t)\,d\zeta = \frac{2}{h}\omega_3(x_1,x_2,t).$$

Here Ψ_α are the rotations of the middle plane around x_α axis, $W^* + \hat{W}$ the vertical deflection of the middle plate, $\Omega_\alpha^* + \hat{\Omega}_\alpha$ the microrotations in the middle plane around x_α axis, Ω_3 the rate of change of the microrotation ϕ_3 along x_3.

Therefore, we can introduce the set of kinematic variables defined as

$$\mathcal{U} = \left[\Psi_\alpha, W^*, \hat{W}, \Omega_\alpha^*, \hat{\Omega}_\alpha, \Omega_3\right],$$

The Cosserat plate stress energy is given as

$$U_P^S = \int_{P_0} \Phi(\mathcal{S},\eta)\,da,$$

where

$$
\Phi(\mathcal{S}, \eta) = -\frac{3\lambda}{h^3\mu(3\lambda + 2\mu)}\left(M_{\alpha\alpha}M_{\beta\beta} + \frac{\alpha + \mu}{2h^3\alpha\mu}3M_{\alpha\beta}^2\right)
$$

$$
+ \frac{3(\alpha - \mu)}{2h^3\alpha\mu}\left(M_{\alpha\beta}\right)\left(M_{\alpha\beta}\right)
$$

$$
+ \frac{\alpha - \mu}{280h^3\alpha\mu}\left[21Q_\alpha\left(5\hat{Q}_\alpha + 4Q_\alpha^*\right)\right]
$$

$$
+ \frac{3(\alpha + \mu)}{160h^3\alpha\mu}\left[8\hat{Q}_\alpha\hat{Q}_\alpha + 15Q_\alpha\hat{Q}_\alpha + 20\hat{Q}_\alpha Q_\alpha^* + 8Q_\alpha^*Q_\alpha^*\right]
$$

$$
- \frac{\gamma - \epsilon}{160h\gamma\epsilon}\left[24R_{\alpha\alpha}^{2*} + 45\hat{R}_{\alpha\alpha} + 60R_{\alpha\beta}^*\hat{R}_{\alpha\beta} + 48R_{12}^*R_{21}^*\right]
$$

$$
+ \frac{\gamma + \epsilon}{160h^3\gamma\epsilon}\left[8R_{\alpha\beta}^{2*} + 15\hat{R}_{\alpha\beta}\hat{R}_{\alpha\beta} + 20R_{\alpha\beta}^*\hat{R}_{\alpha\beta}\right]
$$

$$
+ \frac{3\beta}{80h\gamma(3\beta + 2\gamma)}\left[8R_{\alpha\alpha}^*R_{\beta\beta}^* + 15\hat{R}_{\alpha\alpha}\hat{R}_{\beta\beta} + 20R_{\alpha\alpha}^*\hat{R}_{\alpha\alpha}\right]
$$

$$
+ \frac{3(\gamma + \epsilon)}{2h^3\gamma\epsilon}S_\alpha^2 - \frac{\lambda(5 + \eta)}{10h\mu(3\lambda + 2\mu)}pM_{\alpha\alpha}
$$

$$
+ \frac{(\lambda + \mu)h}{840\mu(3\lambda + 2\mu)}\left(35 + 14\eta + 2\eta^2\right)p^2
$$

$$
+ \frac{(\lambda + \mu)h}{2\mu(3\lambda + 2\mu)}\sigma_0^2,
$$

and

$$
p^* = \eta p(x_1, x_2, t),
$$
$$
\hat{p} = (1 - \eta)p(x_1, x_2, t).
$$

The plate kinetic energy is given as

$$
T_K^\mathcal{U} = \int_{P_0}\Upsilon\left(\frac{\partial\mathcal{U}}{\partial t}\right)da,
$$

where

$$
\Upsilon\left(\frac{\partial\mathcal{U}}{\partial t}\right) = \frac{h^3\rho}{24}\left(\frac{\partial\Psi_\alpha}{\partial t}\right)^2 + \frac{5h\rho}{12}\left(\frac{\partial W^*}{\partial t}\right)^2
$$

$$
+ \frac{h\rho}{2}\left(\frac{\partial\hat{W}}{\partial t}\right)^2 + \frac{5hJ_{\alpha\beta}}{12}\left(\frac{\partial\Omega_\alpha^*}{\partial t}\right)^2
$$

$$
+ \frac{hJ_{\alpha\beta}}{2}\left(\frac{\partial\hat{\Omega}_\alpha}{\partial t}\right)^2 + \frac{h^3J_{33}}{24}\left(\frac{\partial\Omega_3}{\partial t}\right)^2.
$$

$\mathcal{S}$, $\mathcal{U}$, and $\mathcal{E}$ are the Cosserat plate stress, displacement, and strain sets

$$\mathcal{S} = \left[M_{\alpha\beta}, Q_\alpha, Q_\alpha^*, \hat{Q}_\alpha, R_{\alpha\beta}^*, \hat{R}_{\alpha\beta}, S_\beta\right],$$

$$\mathcal{E} = \left[e_{\alpha\beta}, \pi_\beta, \pi_\alpha^*, \hat{\pi}_\alpha, \tau_{\alpha\beta}^*, \hat{\tau}_{\alpha\beta}, \tau_{3\alpha}\right],$$

$$\mathcal{U} = \left[\Psi_\alpha, W^*, \hat{W}, \Omega_\alpha^*, \hat{\Omega}_\alpha, \Omega_3\right].$$

Here we define the variables

$$M_{\alpha\beta} = \frac{h^2}{12}\int_{-1}^{1} m_{\alpha\beta} d\zeta = \frac{h^2}{6} m_{\alpha\beta},$$

$$Q_\alpha = \frac{h}{2}\int_{-1}^{1} q_\alpha\left(1 - \zeta^2\right) d\zeta = \frac{2h}{3} q_\alpha,$$

$$Q_\alpha^* = \frac{h}{2}\int_{-1}^{1} q_\alpha^*\left(1 - \zeta^2\right) d\zeta = \frac{2h}{3} q_\alpha^*,$$

$$\hat{Q}_\alpha = \frac{h}{2}\int_{-1}^{1} \hat{q}_\alpha\left(1 - \zeta^2\right) d\zeta = \frac{2h}{3} \hat{q}_\alpha,$$

$$R_{\alpha\beta}^* = \frac{h}{2}\int_{-1}^{1} r_{\alpha\beta}^*\left(1 - \zeta^2\right) d\zeta = \frac{2h}{3} r_{\alpha\beta}^*,$$

$$\hat{R}_{\alpha\beta} = \frac{h}{2}\int_{-1}^{1} \hat{r}_{\alpha\beta}\left(1 - \zeta^2\right) d\zeta = \frac{2h}{3} \hat{r}_{\alpha\beta},$$

$$S_\alpha = \frac{h^2}{12}\int_{-1}^{1} s_\alpha d\zeta = \frac{h^2}{6} s_\alpha.$$

The following quantities are defined per unit length:

$$M_{11}, M_{22} - \text{bending moments,}$$

$$M_{12}, M_{21} - \text{twisting moments,}$$

$$Q_\alpha - \text{shear forces,}$$

$$Q_\alpha^*, \hat{Q}_\alpha - \text{transverse shear forces,}$$

$$R_{11}^*, R_{22}^*, \hat{R}_{11}, \hat{R}_{22} - \text{Cosserat bending moments,}$$

$$R_{12}^*, R_{21}^*, \hat{R}_{12}, \hat{R}_{21} - \text{Cosserat twisting moments,}$$

$$S_\alpha - \text{Cosserat couple moments.}$$

The dynamic equilibrium system of equations is obtained by substituting the assumptions on stress and couple stress (4.22)–(4.29) and the kinematic assumptions (4.34)–(4.37) into the three-dimensional equilibrium equations (4.1) and (4.2):

$$M_{\alpha\beta,\alpha} - Q_\beta = \rho_1 \frac{\partial^2 \Psi_\beta}{\partial t^2}, \tag{4.45}$$

$$Q^*_{\alpha,\alpha} + p^* = \rho_2 \frac{\partial^2 W^*}{\partial t^2}, \tag{4.46}$$

$$\hat{Q}_{\alpha,\alpha} + \hat{p} = \rho_3 \frac{\partial^2 \hat{W}}{\partial t^2}, \tag{4.47}$$

$$R^*_{\alpha\beta,\alpha} + \varepsilon_{3\beta\gamma}\left(Q^*_\gamma - Q_\gamma\right) = I^*_{\beta\alpha} \frac{\partial^2 \Omega^*_\alpha}{\partial t^2}, \tag{4.48}$$

$$\hat{R}_{\alpha\beta,\alpha} + \varepsilon_{3\beta\gamma}\hat{Q}_\gamma = \hat{I}_{\beta\alpha} \frac{\partial^2 \hat{\Omega}_\alpha}{\partial t^2}, \tag{4.49}$$

$$S_{\alpha,\alpha} + \varepsilon_{3\beta\gamma}M_{\beta\gamma} = I_3 \frac{\partial^2 \Omega_3}{\partial t^2}, \tag{4.50}$$

where

$$\rho_1 = \frac{h^3}{12}\rho,$$

$$\rho_2 = \frac{5h}{6}\rho,$$

$$\rho_3 = \frac{2h}{3}\rho,$$

$$I^*_{\alpha\beta} = \frac{5h}{6}J_{\alpha\beta},$$

$$\hat{I}_{\alpha\beta} = \frac{2h}{3}J_{\alpha\beta},$$

$$I_3 = \frac{h^3}{12}J_{33},$$

and

$$p^*_1 = \eta p(x_1, x_2, t),$$

$$\hat{p}_1 = \frac{2}{3}(1-\eta)p(x_1, x_2, t),$$

with the traction boundary conditions

$$M_{\alpha\beta}n_\beta = M_{o\alpha} \text{ on } \Gamma_\sigma, \tag{4.51}$$

$$Q_\alpha^* n_\alpha = Q_o^* \text{ on } \Gamma_\sigma, \tag{4.52}$$

$$\hat{Q}_\alpha n_\alpha = \hat{Q}_o \text{ on } \Gamma_\sigma, \tag{4.53}$$

$$R_{\alpha\beta}^* n_\beta = R_{o\alpha}^* \text{ on } \Gamma_\sigma, \tag{4.54}$$

$$\hat{R}_{\alpha\beta}n_\beta = \hat{R}_{o\alpha} \text{ on } \Gamma_\sigma, \tag{4.55}$$

$$S_\alpha n_\alpha = S_o \text{ on } \Gamma_\sigma, \tag{4.56}$$

and the displacement boundary conditions

$$\Psi_\alpha = \Psi_{o\alpha} \text{ on } \Gamma_u, \tag{4.57}$$

$$W^* = W_o^* \text{ on } \Gamma_u, \tag{4.58}$$

$$\hat{W} = \hat{W}_o \text{ on } \Gamma_u, \tag{4.59}$$

$$\Omega_\alpha^* = \Omega_{o\alpha}^* \text{ on } \Gamma_u, \tag{4.60}$$

$$\hat{\Omega}_\alpha = \hat{\Omega}_{o\alpha} \text{ on } \Gamma_u, \tag{4.61}$$

$$\Omega_3 = \Omega_{o3} \text{ on } \Gamma_u, \tag{4.62}$$

Constitutive formulas in the reverse form[1]:

$$
\begin{aligned}
M_{\alpha\alpha} =& \frac{h^3\mu(\lambda+\mu)}{3(\lambda+2\mu)}\Psi_{\alpha,\alpha} + \frac{\lambda\mu h^3}{6(\lambda+2\mu)}\Psi_{\beta,\beta} \\
& + \frac{(6p^*+5\hat{p})\lambda h^2}{60(\lambda+2\mu)},
\end{aligned}
\tag{4.63}
$$

$$
\begin{aligned}
M_{\beta\alpha} =& \frac{(\mu-\alpha)h^3}{12}\Psi_{\alpha,\beta} + \frac{h^3(\alpha+\mu)}{12}\Psi_{\beta,\alpha} + \\
& (-1)^\beta \frac{\alpha h^3}{6}\Omega_3,
\end{aligned}
\tag{4.64}
$$

$$
\begin{aligned}
Q_\alpha =& \frac{5h(\alpha+\mu)}{6}\Psi_\alpha + \frac{5(\mu-\alpha)h}{6}\hat{W}_{,\alpha} \\
& + \frac{5(\mu-\alpha)h}{6}W_{,\alpha}^* + (-1)^\beta \frac{5h\alpha}{3}\Omega_\beta^*
\end{aligned}
\tag{4.65}
$$

$$
+ (-1)^\beta \frac{5h\alpha}{3}\hat{\Omega}_\beta,
\tag{4.66}
$$

---

[1] In the following formulas a subindex $\beta = 1$ if $\alpha = 2$ and $\beta = 2$ if $\alpha = 1$.

$$Q_\alpha^* = \frac{5(\mu-\alpha)h}{6}\Psi_\alpha + \frac{5(\mu-\alpha)^2 h}{6(\mu+\alpha)}\hat{W}_{,\alpha}$$
$$+ \frac{5(\mu+\alpha)h}{6}W_{,\alpha}^* + (-1)^\alpha \frac{5h\alpha}{3}\Omega_\beta^*$$
$$+ (-1)^\alpha \frac{5h\alpha(\mu-\alpha)}{3(\mu+\alpha)}\hat{\Omega}_\beta, \qquad (4.67)$$

$$\hat{Q}_\alpha = \frac{8\alpha\mu h}{3(\mu+\alpha)}\hat{W}_{,\alpha} + (-1)^\alpha \frac{8\alpha\mu h}{3(\mu+\alpha)}\hat{\Omega}_\beta, \qquad (4.68)$$

$$R_{\alpha\alpha}^* = \frac{10h\gamma(\beta+\gamma)}{3(\beta+2\gamma)}\Omega_{\alpha,\alpha}^* + \frac{5h\beta\gamma}{3(\beta+2\gamma)}\Omega_{\beta,\beta}^*, \qquad (4.69)$$

$$R_{\beta\alpha}^* = \frac{5(\gamma-\epsilon)h}{6}\Omega_{\beta,\alpha}^* + \frac{5h(\gamma+\epsilon)}{6}\Omega_{\alpha,\beta}^*, \qquad (4.70)$$

$$\hat{R}_{\alpha\alpha} = \frac{8\gamma(\gamma+\beta)h}{3(\beta+2\gamma)}\hat{\Omega}_{\alpha,\alpha} + \frac{4\gamma\beta h}{3(\beta+2\gamma)}\hat{\Omega}_{\beta,\beta}, \qquad (4.71)$$

$$\hat{R}_{\beta\alpha} = \frac{2(\gamma-\epsilon)h}{3}\hat{\Omega}_{\beta,\alpha} + \frac{2(\gamma+\epsilon)h}{3}\hat{\Omega}_{\alpha,\beta}, \qquad (4.72)$$

$$S_\alpha = \frac{\gamma\epsilon h^3}{3(\gamma+\epsilon)}\Omega_{3,\alpha}. \qquad (4.73)$$

and the optimal value η_{opt} of the splitting parameter is given as

$$\eta_{opt} = \frac{2\mathcal{W}^{(00)} - \mathcal{W}^{(10)} - \mathcal{W}^{(01)}}{2\left(\mathcal{W}^{(11)} + \mathcal{W}^{(00)} - \mathcal{W}^{(10)} - \mathcal{W}^{(01)}\right)}, \qquad (4.74)$$

where

$$\mathcal{W}^{(00)} = \mathcal{S}^{(0)} \cdot \mathcal{E}^{(0)}, \qquad (4.75)$$
$$\mathcal{W}^{(01)} = \mathcal{S}^{(0)} \cdot \mathcal{E}^{(1)}, \qquad (4.76)$$
$$\mathcal{W}^{(10)} = \mathcal{S}^{(1)} \cdot \mathcal{E}^{(0)}, \qquad (4.77)$$
$$\mathcal{W}^{(11)} = \mathcal{S}^{(1)} \cdot \mathcal{E}^{(1)}. \qquad (4.78)$$

We also assume that the initial condition can be presented in the form

$$\mathcal{U}(x_1, x_2, 0) = \mathcal{U}^0(x_1, x_2),$$
$$\frac{\partial \mathcal{V}}{\partial t}(x_1, x_2, 0) = \mathcal{V}^0(x_1, x_2).$$

Cosserat Plate Dynamic Equilibrium Equations

$$M_{\alpha\beta,\alpha} - Q_\beta = \rho_1 \frac{\partial^2 \Psi_\beta}{\partial t^2}$$

$$Q^*_{\alpha,\alpha} + p^*_1 = \rho_2 \frac{\partial^2 W^*}{\partial t^2}$$

$$\hat{Q}_{\alpha,\alpha} + \hat{p}_1 = \rho_3 \frac{\partial^2 \hat{W}}{\partial t^2}$$

$$R^*_{\alpha\beta,\alpha} + \varepsilon_{3\beta\gamma}\left(Q^*_\gamma - Q_\gamma\right) = I^*_{\beta\alpha} \frac{\partial^2 \Omega^*_\alpha}{\partial t^2}$$

$$\hat{R}_{\alpha\beta,\alpha} + \varepsilon_{3\beta\gamma}\hat{Q}_\gamma = \hat{I}_{\beta\alpha} \frac{\partial^2 \hat{\Omega}_\alpha}{\partial t^2}$$

$$S_{\alpha,\alpha} + \varepsilon_{3\beta\gamma}M_{\beta\gamma} = I_3 \frac{\partial^2 \Omega_3}{\partial t^2}$$

4.3 DYNAMIC FIELD EQUATIONS

The Cosserat plate field equations are obtained by substituting the relations (3.82)–(3.92) into the system of equations (4.45)–(4.50):

$$L\,\mathcal{U} = K \frac{\partial^2 \mathcal{U}}{\partial t^2} + f(\eta), \qquad (4.79)$$

where the operator L here is given as

$$L = \begin{bmatrix}
L_{11} & L_{12} & L_{13} & L_{14} & 0 & L_{16} & 0 & L_{18} & L_{19} \\
L_{12} & L_{22} & L_{23} & L_{24} & L_{25} & 0 & L_{27} & 0 & L_{29} \\
L_{31} & L_{32} & L_{33} & L_{34} & L_{35} & L_{36} & L_{37} & L_{38} & 0 \\
L_{41} & L_{42} & L_{43} & L_{44} & L_{45} & L_{46} & L_{47} & L_{48} & 0 \\
0 & L_{52} & L_{53} & L_{54} & L_{55} & L_{56} & L_{57} & 0 & 0 \\
L_{61} & 0 & L_{63} & L_{64} & L_{65} & L_{66} & 0 & L_{68} & 0 \\
0 & 0 & 0 & L_{74} & 0 & 0 & L_{77} & L_{78} & 0 \\
0 & 0 & 0 & L_{84} & 0 & 0 & L_{87} & L_{88} & 0 \\
L_{91} & L_{92} & 0 & 0 & 0 & 0 & 0 & 0 & L_{99}
\end{bmatrix},$$

the solution vector v of the kinematic variables is defined as

$$\mathcal{U} = \left[\Psi_1, \Psi_2, W^*, \hat{W}, \Omega_1^*, \Omega_2^*, \hat{\Omega}_1, \hat{\Omega}_2, \Omega_3\right]^T,$$

the matrix K is given as

$$K = \begin{bmatrix}
\frac{h^3}{12}\rho & 0 & 0 & 0 & 0 & 0 & 0 & 0 & 0 \\
0 & \frac{h^3}{12}\rho & 0 & 0 & 0 & 0 & 0 & 0 & 0 \\
0 & 0 & \frac{5h}{6}\rho & \frac{5h}{6}\rho & 0 & 0 & 0 & 0 & 0 \\
0 & 0 & 0 & \frac{5h}{6}\rho & 0 & 0 & 0 & 0 & 0 \\
0 & 0 & 0 & 0 & \frac{5h}{6}J_{11} & \frac{5h}{6}J_{12} & 0 & 0 & 0 \\
0 & 0 & 0 & 0 & \frac{5h}{6}J_{21} & \frac{5h}{6}J_{22} & 0 & 0 & 0 \\
0 & 0 & 0 & 0 & 0 & 0 & \frac{5h}{6}J_{11} & \frac{5h}{6}J_{12} & 0 \\
0 & 0 & 0 & 0 & 0 & 0 & \frac{5h}{6}J_{21} & \frac{5h}{6}J_{22} & 0 \\
0 & 0 & 0 & 0 & 0 & 0 & 0 & 0 & \frac{h^3}{12}J_{33}
\end{bmatrix},$$

and the right-hand side $f(\eta)$ vector is

$$f(\eta) = \begin{bmatrix}
-\frac{h^3\lambda\left(6p_{,1}^* + 5\hat{p}_{,1}\right)}{120(\lambda+2\mu)} \\
-\frac{h^3\lambda\left(6p_{,2}^* + 5\hat{p}_{,2}\right)}{120(\lambda+2\mu)} \\
-\frac{6p^* + 5\hat{p}}{6} \\
-p^* \\
0 \\
0 \\
0 \\
0 \\
0
\end{bmatrix},$$

where the pressures p^* and $\hat{p}$ are defined as before:

$$p^*(x_1, x_2, t) = \eta\, p(x_1, x_2, t), \tag{4.80}$$
$$\hat{p}(x_1, x_2, t) = (1-\eta)\, p(x_1, x_2, t). \tag{4.81}$$

The operators L_{ij} are defined as follows:

$$L_{11} = c_1 \frac{\partial^2}{\partial x_1^2} + c_2 \frac{\partial^2}{\partial x_2^2} - c_3, \qquad L_{12} = c_4 \frac{\partial^2}{\partial x_1 x_2},$$

$$L_{13} = c_5 \frac{\partial}{\partial x_1}, \qquad L_{14} = c_5 \frac{\partial}{\partial x_1},$$

$$L_{16} = c_6, \qquad L_{18} = c_6,$$

$$L_{19} = c_7 \frac{\partial}{\partial x_2}, \qquad L_{21} = c_4 \frac{\partial^2}{\partial x_1 x_2},$$

$$L_{22} = c_2 \frac{\partial^2}{\partial x_1^2} + c_1 \frac{\partial^2}{\partial x_2^2} - c_3, \qquad L_{23} = c_5 \frac{\partial}{\partial x_2},$$

$$L_{24} = c_5 \frac{\partial}{\partial x_2}, \qquad L_{25} = -c_6,$$

$$L_{27} = -c_6, \qquad L_{29} = -c_7 \frac{\partial}{\partial x_1},$$

$$L_{31} = c_5 \frac{\partial}{\partial x_1}, \qquad L_{32} = -c_5 \frac{\partial}{\partial x_1},$$

$$L_{33} = c_3 \frac{\partial^2}{\partial x_1^2} + c_3 \frac{\partial^2}{\partial x_2^2}, \qquad L_{34} = c_3 \frac{\partial^2}{\partial x_1^2} + c_3 \frac{\partial^2}{\partial x_2^2},$$

$$L_{35} = -c_6 \frac{\partial}{\partial x_2}, \qquad L_{36} = c_6 \frac{\partial}{\partial x_1},$$

$$L_{37} = -c_6 \frac{\partial}{\partial x_2}, \qquad L_{38} = c_6 \frac{\partial}{\partial x_1},$$

$$L_{41} = c_5 \frac{\partial}{\partial x_1}, \qquad L_{42} = -c_5 \frac{\partial}{\partial x_1},$$

$$L_{43} = c_3 \frac{\partial^2}{\partial x_1^2} + c_3 \frac{\partial^2}{\partial x_2^2}, \qquad L_{44} = c_8 \frac{\partial^2}{\partial x_1^2} + c_8 \frac{\partial^2}{\partial x_2^2},$$

$$L_{45} = -c_6 \frac{\partial}{\partial x_2}, \qquad L_{46} = c_6 \frac{\partial}{\partial x_1},$$

$$L_{47} = -c_9 \frac{\partial}{\partial x_2}, \qquad L_{48} = c_9 \frac{\partial}{\partial x_1},$$

$$L_{52} = -c_6, \qquad L_{53} = c_6 \frac{\partial}{\partial x_2},$$

$$L_{54} = c_9 \frac{\partial}{\partial x_2}, \qquad L_{55} = c_{10} \frac{\partial^2}{\partial x_1^2} + c_{11} \frac{\partial^2}{\partial x_2^2} - 2c_6,$$

$$L_{56} = c_{12} \frac{\partial^2}{\partial x_1 x_2}, \qquad L_{57} = -c_{13},$$

$$L_{61} = c_6, \qquad L_{63} = -c_6 \frac{\partial}{\partial x_2},$$

$$L_{64} = -c_9 \frac{\partial}{\partial x_2}, \qquad L_{65} = c_{12} \frac{\partial^2}{\partial x_1 x_2},$$

$$L_{66} = c_{10} \frac{\partial^2}{\partial x_1^2} + c_{11} \frac{\partial^2}{\partial x_2^2} - 2c_6, \qquad L_{67} = -c_{13},$$

$$L_{74} = c_{14} \frac{\partial}{\partial x_2}, \qquad L_{77} = c_{10} \frac{\partial^2}{\partial x_1^2} + c_{11} \frac{\partial^2}{\partial x_2^2} - c_{14},$$

$$L_{78} = c_{12} \frac{\partial^2}{\partial x_1 x_2}, \qquad L_{84} = c_{14} \frac{\partial}{\partial x_2},$$

$$L_{78} = c_{12} \frac{\partial^2}{\partial x_1 x_2}, \qquad L_{88} = c_{11} \frac{\partial^2}{\partial x_1^2} + c_{10} \frac{\partial^2}{\partial x_2^2} - c_{14},$$

$$L_{91} = -c_7 \frac{\partial}{\partial x_2}, \qquad L_{92} = c_7 \frac{\partial}{\partial x_1},$$

$$L_{99} = c_{15} \frac{\partial^2}{\partial x_1^2} + c_{15} \frac{\partial^2}{\partial x_2^2} - 2c_7,$$

where c_i are the constants given as

$$c_1 = \frac{h^4 \mu (\lambda + \mu)}{6(\lambda + 2\mu)}, \qquad c_2 = \frac{h^4 (\alpha + \mu)}{24},$$

$$c_3 = \frac{5h(\alpha + \mu)}{6}, \qquad c_4 = \frac{h^4 (\mu(3\lambda + 2\mu) - \alpha(\lambda + 2\mu))}{24(\lambda + 2\mu)},$$

$$c_5 = \frac{5h(\alpha - \mu)}{6}, \qquad c_6 = \frac{5h\alpha}{3},$$

$$c_7 = \frac{h^4 \alpha}{12}, \qquad c_8 = \frac{5h(\alpha - \mu)^2}{6(\alpha + \mu)},$$

$$c_9 = \frac{5h\alpha(\alpha - \mu)}{3(\alpha + \mu)}, \qquad c_{10} = \frac{10h\gamma(\beta + \gamma)}{3(\beta + 2\gamma)},$$

$$c_{11} = \frac{5h(\gamma + \epsilon)}{6}, \qquad c_{12} = \frac{5h(2\gamma(\gamma - \epsilon) + \beta(3\gamma - \epsilon))}{6(\beta + 2\gamma)},$$

$$c_{13} = \frac{10h\alpha^2}{3(\alpha + \mu)}, \qquad c_{14} = \frac{10h\alpha\mu}{3(\alpha + \mu)},$$

$$c_{15} = \frac{h^4 \gamma \epsilon}{6(\gamma + \epsilon)}.$$

CHAPTER 5

Validation of the Cosserat Plate Theory

5.1 SCOPE OF THE CHAPTER

This chapter provides the validation of the Cosserat Plate Theory. We provide the solutions for the simply supported plate and compare them with the exact solutions of the three-dimensional Cosserat Elasticity. The comparison shows the consilience between the Cosserat Plate Theory and the three-dimensional Cosserat Elasticity.

5.2 COSSERAT SAMPLE PLATE

We will consider a Cosserat body B being a rectangular cuboid of size $[0,a] \times [0,a] \times \left[-\frac{h}{2}, \frac{h}{2}\right]$, where a and h are positive real numbers. Let the sets T and B be the top and the bottom surfaces contained in the planes $x_3 = \frac{h}{2}$ and $x_3 = -\frac{h}{2}$ respectively, and the curve Γ be the lateral part of the boundary. Let us define the boundaries Γ_1,

DOI: 10.1201/9781003190264-5

Γ_2, Γ_3, and Γ_4 as

$$\Gamma_1 = \{(x_1, x_2, x_3) : x_1 = 0, x_2 \in [0, a]\},$$
$$\Gamma_2 = \{(x_1, x_2, x_3) : x_1 = a, x_2 \in [0, a]\},$$
$$\Gamma_3 = \{(x_1, x_2, x_3) : x_1 \in [0, a], x_2 = 0\},$$
$$\Gamma_4 = \{(x_1, x_2, x_3) : x_1 \in [0, a], x_2 = a\},$$

where $\Gamma_1 \cup \Gamma_2 \cup \Gamma_3 \cup \Gamma_4 = \Gamma$.

The Cosserat body B can be also viewed as a square Cosserat plate P of size $[0, a] \times [0, a]$, having the thickness h and the boundary $G = G_1 \cup G_2 \cup G_3 \cup G_4$, where

$$G_1 = \{(x_1, x_2) : x_1 = 0, x_2 \in [0, a]\},$$

$$G_2 = \{(x_1, x_2) : x_1 = a, x_2 \in [0, a]\},$$

$$[6pt]G_3 = \{(x_1, x_2) : x_1 \in [0, a], x_2 = 0\},$$

$$G_4 = \{(x_1, x_2) : x_1 \in [0, a], x_2 = a\}.$$

We will consider the simply supported boundary conditions for the Cosserat plate, which are given in the mixed Dirichlet-Neumann form in Table 5.1. For the comparison purposes we will consider the thickness of the plate $h = 0.1$m with the width-to-thickness ratio a/h varying from 5 to 30 (see Figure 5.1).

The plate is assumed to be made of the polyurethane foam with the following values of the Lamé and Cosserat Elasticity parameters:

$$\lambda = 762.616,$$
$$\mu = 103.993,$$
$$\alpha = 4.333,$$
$$\beta = 39.975,$$
$$\gamma = 39.975,$$
$$\epsilon = 4.505.$$

We consider a low density rigid foam characterized by the density value $\rho = 34$ kg/m^3 and the microinertia tensor $\mathbf{J}$ given as

$$\mathbf{J} = \begin{pmatrix} J_x & 0 & 0 \\ 0 & J_y & 0 \\ 0 & 0 & J_z \end{pmatrix}.$$

TABLE 5.1 Simply supported boundary conditions for the Cosserat plate.

Boundary $G_1 \cup G_2$	Boundary $G_3 \cup G_4$
$\frac{\partial \Psi_1}{\partial n} = 0$	$\frac{\partial \Psi_2}{\partial n} = 0$
$\Psi_2 = 0$	$\Psi_1 = 0$
$W^* = 0$	$W^* = 0$
$\hat{W} = 0$	$\hat{W} = 0$
$\Omega_1^* = 0$	$\Omega_2^* = 0$
$\frac{\partial \Omega_2^*}{\partial n} = 0$	$\frac{\partial \Omega_1^*}{\partial n} = 0$
$\hat{\Omega}_1 = 0$	$\hat{\Omega}_2 = 0$
$\frac{\partial \hat{\Omega}_2}{\partial n} = 0$	$\frac{\partial \hat{\Omega}_1}{\partial n} = 0$
$\Omega_3 = 0$	$\Omega_3 = 0$

Here, the principal moments of microinertia of the microelements $J_x = J_y = J_z = 0.001$ are assumed to be constant throughout the Cosserat plate P.

5.3 VALIDATION OF THE COSSERAT PLATE STATICS

The distribution of the pressure is assumed to have a sinusoidal form:

$$p(x_1, x_2) = \sin\left(\frac{\pi x_1}{a}\right) \cdot \sin\left(\frac{\pi x_2}{a}\right).$$

We will solve the two-dimensional system of Cosserat plate field equations (3.93) by applying the method of separation of variables similar to [49]. As a result we obtain the kinematic variables in the following form:

$$\Psi_1 = A_1(\eta) \cdot \cos\left(\frac{\pi x_1}{a}\right) \cdot \sin\left(\frac{\pi x_2}{a}\right), \tag{5.1}$$

$$\Psi_2 = A_2(\eta) \cdot \sin\left(\frac{\pi x_1}{a}\right) \cdot \cos\left(\frac{\pi x_2}{a}\right), \tag{5.2}$$

$$W^* = A_3(\eta) \cdot \sin\left(\frac{\pi x_1}{a}\right) \cdot \sin\left(\frac{\pi x_2}{a}\right), \tag{5.3}$$

$$\hat{W} = A_4(\eta) \cdot \sin\left(\frac{\pi x_1}{a}\right) \cdot \sin\left(\frac{\pi x_2}{a}\right), \tag{5.4}$$

$$\Omega_1^* = A_5(\eta) \cdot \sin\left(\frac{\pi x_1}{a}\right) \cdot \cos\left(\frac{\pi x_2}{a}\right), \tag{5.5}$$

$$\Omega_2^* = A_6(\eta) \cdot \cos\left(\frac{\pi x_1}{a}\right) \cdot \sin\left(\frac{\pi x_2}{a}\right), \tag{5.6}$$

$$\hat{\Omega}_1 = A_7(\eta) \cdot \sin\left(\frac{\pi x_1}{a}\right) \cdot \cos\left(\frac{\pi x_2}{a}\right), \tag{5.7}$$

$$\hat{\Omega}_2 = A_8(\eta) \cdot \cos\left(\frac{\pi x_1}{a}\right) \cdot \sin\left(\frac{\pi x_2}{a}\right), \tag{5.8}$$

$$\Omega_3 = A_9(\eta) \cdot \cos\left(\frac{\pi x_1}{a}\right) \cdot \cos\left(\frac{\pi x_2}{a}\right), \tag{5.9}$$

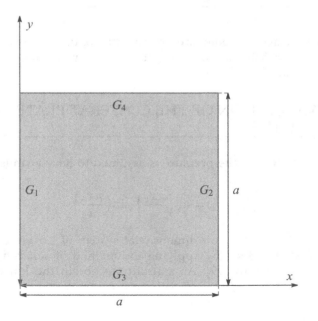

Figure 5.1 Sample Cosserat Plate.

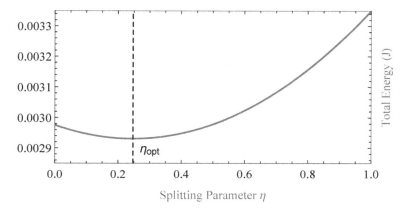

Figure 5.2 Total energy of the plate as a function of the splitting parameter η. The optimal value $\eta = 0.249$ minimizes the energy for the plate (width-to-thickness ratio $a/h = 10$).

where $A_i(\eta)$ ($i = 1,...,9$) are real valued functions of the splitting parameter η. We find $A_i(\eta)$ by substituting the expressions (5.1)–(5.9) into the system of equations (3.93) and solving the obtained system of 9 linear equations for $A_i(\eta)$.

The numerical values for $A_i(\eta)$ correspond to the optimal value of the splitting parameter η_{opt}, which minimizes the plate stress energy (3.42). We consider the optimal value of the splitting parameter that minimizes the stress energy with the relative error of 10^{-4}. The graph in Figure 5.2 represents the total energy of the plate as a function of the splitting parameter η. It shows the optimal value $\eta_{opt} = 0.249$, which minimizes the plate stress energy for the square Cosserat plate with the width-to-thickness ratio $a/h = 10$.

Figures 5.3–5.6 represent the qualitative comparisons of the displacements u_i and the microrotations φ_i obtained from the

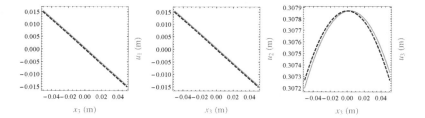

Figure 5.3 Comparison of the displacements u_i.

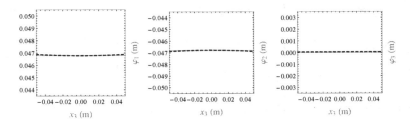

Figure 5.4 Comparison of the microrotations ϕ_i.

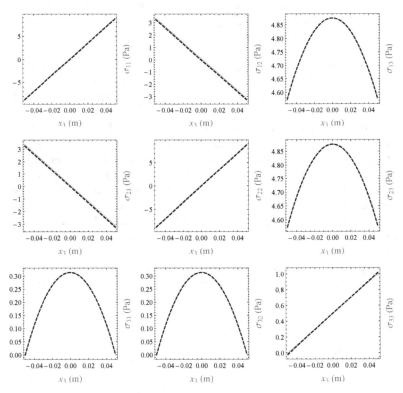

Figure 5.5 Comparison of the stresses σ_{ij}.

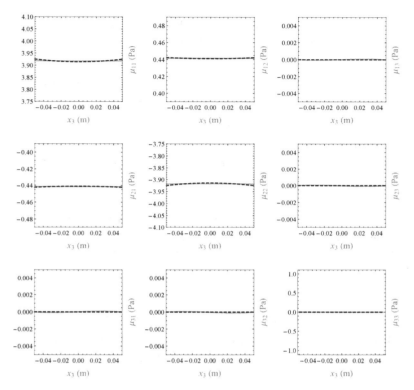

Figure 5.6   Comparison of the couple stresses μ_{ij}.

Cosserat Plate Theory, and the exact solutions of the three-dimensional Cosserat Elasticity found in Chapter 2.

Tables 5.2–5.7 represent the numerical comparisons of the vertical deflection u_3, shear displacements u_α, and microrotations φ_α obtained from the Cosserat Plate Theory and the exact solutions of the three-dimensional Cosserat Elasticity found in Chapter 2. For the plates with the width-to-thickness ratio greater or equal to 5, the kinematic variables have a relative error of order 1% in comparison with the exact three-dimensional solutions. This shows the consilience between the Cosserat Plate Theory and the three-dimensional Cosserat Elasticity.

TABLE 5.2 Comparison of the vertical deflection u_3 (m) for the values of the width-to-thickness ratio a/h = 5, 10, 15.

a/h	5	10	15
Cosserat Plate Theory	0.00374561	0.02374524	0.06242873
Cosserat 3D Elasticity	0.00374336	0.02374431	0.06242824
Relative Error	**0.06013%**	**0.00389%**	**0.00078%**

TABLE 5.3 Comparison of the vertical deflection u_3 (m) for the values of the width-to-thickness ratio a/h = 20, 25, 30.

a/h	20	25	30
Cosserat Plate Model	0.12076297	0.20133625	0.30787104
Cosserat 3D Elasticity	0.12076264	0.20133599	0.30787077
Relative Error	**0.00026%**	**0.00013%**	**0.00008%**

TABLE 5.4 Comparison of displacements u_α (m) for the values of the width-to-thickness ratio a/h = 5, 10, 15.

a/h	5	10	15
Cosserat Plate Model	0.00008632	0.00062733	0.00205503
Cosserat 3D Elasticity	0.00008525	0.00062141	0.00204337
Relative Error	**1.25513%**	**0.95267%**	**0.57062%**

TABLE 5.5 Comparison of displacements u_α (m) for the values of the width-to-thickness ratio a/h = 20, 25, 30.

a/h	20	25	30
Cosserat Plate Model	0.00481496	0.00935158	0.01610936
Cosserat 3D Elasticity	0.00479483	0.00931940	0.01606072
Relative Error	**0.41982%**	**0.34530%**	**0.30285%**

TABLE 5.6 Comparison of microrotations φ_α for the values of the width-to-thickness ratio a/h = 5, 10, 15

a/h	5	10	15
Cosserat Plate Model	0.00009767	0.00133873	0.00526100
Cosserat 3D Elasticity	0.00009887	0.00134307	0.00526864
Relative Error	**1.21371%**	**0.32314%**	**0.14500%**

TABLE 5.7 Comparison of microrotations φ_α for the values of the width-to-thickness ratio a/h = 20, 25, 30

a/h	20	25	30
Cosserat Plate Model	0.01324364	0.02663502	0.04678241
Cosserat 3D Elasticity	0.01325449	0.02664900	0.04679946
Relative Error	**0.08185%**	**0.05245%**	**0.03643%**

5.4 VALIDATION OF THE COSSERAT PLATE DYNAMICS

The distribution of the pressure is assumed to have a sinusoidal form:

$$p(x_1, x_2, t) = \sin\left(\frac{\pi x_1}{a}\right) \cdot \sin\left(\frac{\pi x_2}{a}\right) \cdot \sin(\omega t).$$

We will solve the eigenvalue problem for the two-dimensional system of Cosserat plate field equations (4.79) by applying the method of separation of variables similar to [49]. As a result we obtain the kinematic variables in the following form:

$$\Psi_1^{(nm)} = A_1 \cdot f_1^{nm}(x_1, x_2, t) + B_1 \cdot f_2^{nm}(x_1, x_2, t), \tag{5.10}$$

$$\Psi_2^{(nm)} = A_2 \cdot f_2^{nm}(x_1, x_2, t) + B_2 \cdot f_1^{nm}(x_1, x_2, t), \tag{5.11}$$

$$W^{*(nm)} = A_3 \cdot f_3^{(nm)}(x_1, x_2, t) + B_3 \cdot f_4^{nm}(x_1, x_2, t), \tag{5.12}$$

$$\hat{W}^{(nm)} = A_4 \cdot f_3^{nm}(x_1, x_2, t) + B_4 \cdot f_4^{nm}(x_1, x_2, t), \tag{5.13}$$

$$\Omega_1^{*(nm)} = A_5 \cdot f_2^{nm}(x_1, x_2, t) + B_5 \cdot f_1^{nm}(x_1, x_2, t), \tag{5.14}$$

$$\Omega_2^{*(nm)} = A_6 \cdot f_1^{nm}(x_1, x_2, t) + B_6 \cdot f_2^{nm}(x_1, x_2, t), \tag{5.15}$$

$$\hat{\Omega}_1^{(nm)} = A_7 \cdot f_2^{nm}(x_1, x_2, t) + B_7 \cdot f_1^{nm}(x_1, x_2, t), \tag{5.16}$$

$$\hat{\Omega}_2^{(nm)} = A_8 \cdot f_1^{nm}(x_1, x_2, t) + B_8 \cdot f_1^{nm}(x_1, x_2, t), \tag{5.17}$$

$$\Omega_3^{(nm)} = A_9 \cdot f_4^{nm}(x_1, x_2, t) + B_9 \cdot f_3^{nm}(x_1, x_2, t), \tag{5.18}$$

where A_i and B_i are constants and the functions $f_i^{nm}(x_1, x_2, t)$ are defined as

$$f_1^{nm}(x_1, x_2, t) = \cos\left(\frac{n\pi x_1}{a}\right) \cdot \sin\left(\frac{m\pi x_2}{a}\right) \cdot \sin(\omega t),$$

$$f_2^{nm}(x_1, x_2, t) = \sin\left(\frac{n\pi x_1}{a}\right) \cdot \cos\left(\frac{m\pi x_2}{a}\right) \cdot \sin(\omega t),$$

$$f_3^{nm}(x_1, x_2, t) = \sin\left(\frac{n\pi x_1}{a}\right) \cdot \sin\left(\frac{m\pi x_2}{a}\right) \cdot \sin(\omega t),$$

$$f_4^{nm}(x_1, x_2, t) = \cos\left(\frac{n\pi x_1}{a}\right) \cdot \cos\left(\frac{m\pi x_2}{a}\right) \cdot \sin(\omega t).$$

We solve the eigenvalue problem by substituting these expressions into the system of equations (4.79) and solving a standard eigenvalue problem for a system of 18 algebraic equations.

TABLE 5.8   Comparison of the Kirchhoff frequency ω_K (Hz) calculated using the Cosserat Plate Theory and the three-dimensional Cosserat Elasticity for the values of the width-to-thickness ratio a/h = 5, 10, 15.

a/h	5	10	15
Cosserat Plate Model	8.96958	3.56063	2.18625
Cosserat 3D Elasticity	8.61476	3.48762	2.16129
Relative Error	3.96%	2.05%	1.14%

Tables 5.8 and 5.9 represent the numerical comparisons of the Kirchhoff frequency ω_K obtained from the Cosserat Plate Theory and the exact eigenfrequencies of the three-dimensional Cosserat Elasticity found in Chapter 1. Tables 5.10 and 5.11 represent the numerical comparisons of the Mindlin-Reissner frequency ω_{MR} obtained from the Cosserat Plate Theory and the exact eigenfrequencies of the three-dimensional Cosserat Elasticity found in Chapter 1. For the plates with the width to thickness ratio greater or equal to 15, the eigenfrequencies have a relative error of order 1% in comparison with the exact three-dimensional solutions.

TABLE 5.9   Comparison of the Kirchhoff frequency ω_K (Hz) calculated using the Cosserat Plate Theory and the three-dimensional Cosserat Elasticity for the values of the width-to-thickness ratio a/h = 20, 25, 30.

a/h	20	25	30
Cosserat Plate Model	1.56808	1.21281	0.97999
Cosserat 3D Elasticity	1.55678	1.20672	0.97631
Relative Error	0.72%	0.50%	0.38%

TABLE 5.10 Comparison of the Mindlin-Reissner frequency ω_{MR} (Hz) calculated using the Cosserat Plate Theory and the three-dimensional Cosserat Elasticity for the values of the width-to-thickness ratio $a/h = 5$, 10, 15.

a/h	5	10	15
Cosserat Plate Model	64.9537	58.7313	57.4417
Cosserat 3D Elasticity	63.5062	58.0750	56.9498
Relative Error	2.23%	1.12%	0.86%

TABLE 5.11 Comparison of the Mindlin-Reissner frequency ω_{MR} (Hz) calculated using the Cosserat Plate Theory and the three-dimensional Cosserat Elasticity for the values of the width-to-thickness ratio $a/h = 20$, 25, 30.

a/h	20	25	30
Cosserat Plate Model	56.9537	56.6994	56.5352
Cosserat 3D Elasticity	56.5259	56.3066	56.1665
Relative Error	0.75%	0.69%	0.65%

This shows the consilience between the Cosserat Plate Theory and the three-dimensional Cosserat Elasticity.

CHAPTER 6

Finite Element Method for Cosserat Plates

6.1   SCOPE OF THE CHAPTER

This chapter presents the Finite Element Method for Cosserat Plates. It consists of the classic Galerkin finite element for the elliptic system of nine partial differential equations and the energy minimization procedure for the calculation of the optimal value of the splitting parameter. We discuss the weak formulation of the bending problem and present the results of the Finite Element computations for the clamped Cosserat plates of different shapes under different loads.

6.2   CLAMPED COSSERAT PLATE BENDING PROBLEM

Let us consider the Cosserat plate P of thickness h having the boundary G. The system of field equations (3.93) represents a system of nine partial differential equations:

$$L\,\mathcal{U} = f\,(\eta),\tag{6.1}$$

DOI: 10.1201/9781003190264-6

where L is an elliptic differential operator that acts on the vector of nine kinematic variables

$$\mathcal{U} = \left[\Psi_1, \Psi_2, W^*, \hat{W}, \Omega_1^*, \Omega_2^*, \hat{\Omega}_1, \hat{\Omega}_2, \Omega_3 \right]$$

and $f(\eta)$ is the right-hand side vector, that in general depends on the splitting parameter η.

Let us consider the clamped boundary conditions

$$\Psi_1 = 0 \text{ on } G, \tag{6.2}$$

$$\Psi_2 = 0 \text{ on } G, \tag{6.3}$$

$$W^* = 0 \text{ on } G, \tag{6.4}$$

$$W = 0 \text{ on } G, \tag{6.5}$$

$$\Omega_1^* = 0 \text{ on } G, \tag{6.6}$$

$$\Omega_2^* = 0 \text{ on } G, \tag{6.7}$$

$$\hat{\Omega}_1 = 0 \text{ on } G, \tag{6.8}$$

$$\hat{\Omega}_2 = 0 \text{ on } G, \tag{6.9}$$

$$\Omega_3 = 0 \text{ on } G, \tag{6.10}$$

which represent the homogeneous Dirichlet boundary conditions.

6.3 FINITE ELEMENT ALGORITHM FOR COSSERAT ELASTIC PLATES

The right-hand side of the system (6.1) depends on the splitting parameter η and so does the solution $\mathcal{U}$. Therefore, the solution of the Cosserat plate bending problem requires not only solving this system of equations, but also an additional energy minimization procedure for the calculation of the value of the splitting parameter. This optimal value of the splitting parameter η_{opt} corresponds to the unique solution. Considering that the elliptic systems of partial differential equations correspond to a state where the minimum of the energy is reached, the optimal value of the splitting parameter η_{opt} should minimize the plate energy (3.42).

The Finite Element Method for Cosserat elastic plates is based on solving the system (3.93) for two different values of the splitting parameter, and then calculating the stresses, strains, the corresponding work, and the optimal value of the splitting parameter. After this the optimal solution can be easily obtained as a linear combination of the two previously found solutions.

As in Chapter 2, let us consider two separate problems:

$$L\mathcal{U}_0 = \mathcal{F}(0), \tag{6.11}$$

$$L\mathcal{U}_1 = \mathcal{F}(1), \tag{6.12}$$

where $\mathcal{U}_0$ and $\mathcal{U}_1$ represent the solutions of the system of field equations for $\eta = 0$ and $\eta = 1$, respectively. Let $\mathcal{S}^{(0)}$ and $\mathcal{E}^{(0)}$ be the stress and strain sets that correspond to the value of $\eta = 0$ and $\mathcal{S}^{(1)}$ and $\mathcal{E}^{(1)}$ be the stress and strain sets that correspond to the value of $\eta = 1$.

The optimal value η_{opt} of the splitting parameter η can be found using the formula developed in Chapter 3:

$$\eta_{\text{opt}} = \frac{2\mathcal{W}^{(00)} - \mathcal{W}^{(10)} - \mathcal{W}^{(01)}}{2\left(\mathcal{W}^{(11)} + \mathcal{W}^{(00)} - \mathcal{W}^{(10)} - \mathcal{W}^{(01)}\right)}, \tag{6.13}$$

where the work $\mathcal{W}^{(00)}$, $\mathcal{W}^{(01)}$, $\mathcal{W}^{(10)}$, and $\mathcal{W}^{(11)}$ is defined as

$$\mathcal{W}^{(00)} = \mathcal{S}^{(0)} \cdot \mathcal{E}^{(0)}, \tag{6.14}$$

$$\mathcal{W}^{(01)} = \mathcal{S}^{(0)} \cdot \mathcal{E}^{(1)}, \tag{6.15}$$

$$\mathcal{W}^{(10)} = \mathcal{S}^{(1)} \cdot \mathcal{E}^{(0)}, \tag{6.16}$$

$$\mathcal{W}^{(11)} = \mathcal{S}^{(1)} \cdot \mathcal{E}^{(1)}. \tag{6.17}$$

The solution $\mathcal{U}$ of the system of field equations (6.1) is found as a linear combination of the solutions $\mathcal{U}_0$ and $\mathcal{U}_1$:

$$\mathcal{U} = \left(1 - \eta_{\text{opt}}\right) \cdot \mathcal{U}_0 + \eta_{\text{opt}} \cdot \mathcal{U}_1. \tag{6.18}$$

Finite Element Algorithm for the Cosserat Plates

1. Solve the systems (6.11) and (6.12) for the sets of kinematic variables $\mathcal{U}_0$ and $\mathcal{U}_1$, respectively.

2. Calculate the components of the Cosserat plate stress sets $\mathcal{S}^{(0)}$ and $\mathcal{S}^{(1)}$ from the sets of the kinematic variables $\mathcal{U}_0$ and $\mathcal{U}_1$, using the constitutive formulas in the reverse form (3.82)–(3.92).

3. Calculate the components of the Cosserat plate strain sets $\mathcal{E}^{(0)}$ and $\mathcal{E}^{(1)}$ from the sets of kinematic variables $\mathcal{U}_0$ and $\mathcal{U}_1$, using the strain-displacement relations (3.48)–(3.54).

4. Find the work $\mathcal{W}^{(00)}$, $\mathcal{W}^{(11)}$, $\mathcal{W}^{(10)}$, and $\mathcal{W}^{(01)}$ by substituting the Cosserat plate stress and strain sets $\mathcal{S}^{(0)}$, $\mathcal{S}^{(1)}$, $\mathcal{E}^{(0)}$, and $\mathcal{E}^{(1)}$ into the definitions (6.14)–(6.17).

5. Substitute the values of the work $\mathcal{W}^{(00)}$, $\mathcal{W}^{(11)}$, $\mathcal{W}^{(10)}$, and $\mathcal{W}^{(01)}$ into the expression for the optimal value of the splitting parameter η_{opt} (6.13).

6. Compute the solution $\mathcal{U}$ of the system of field equations (6.1) as a linear combinations of the sets of kinematic variables $\mathcal{U}_0$ and $\mathcal{U}_1$ using the formula (6.18).

6.4 SOLUTIONS SPACE FOR THE CLAMPED COSSERAT PLATE

In this section, we will define the Hilbert space for the solutions of the clamped Cosserat plate bending problem. Let us denote by $\mathbf{L}^2(P)$ the standard space of square-integrable functions defined everywhere on P:

$$\mathbf{L}^2(P) = \left\{ v : \int_P v^2 ds < \infty \right\},$$

and by $\mathbf{H}^1(P)$ the Hilbert space of functions that are square-integrable together with their first partial derivatives:

$$\mathbf{H}^1(P) = \left\{ v : v \in L^2(P), \partial_i v \in L^2(P) \right\}.$$

Let us denote the Hilbert space of functions from $\mathbf{H}^1(P)$ that vanish on the boundary as in [9]:

$$\mathbf{H}_0^1(P) = \left\{ v \in \mathbf{H}^1(P), v = 0 \text{ on } \partial P \right\}.$$

The space $\mathbf{H}_0^1(P)$ is equipped with the inner product:

$$\langle u, v \rangle_{\mathbf{H}_0^1} = \int_P (uv + \partial_i u \partial_i v) \, ds \text{ for } u, v \in \mathbf{H}_0^1(P).$$

In case of the clamped Cosserat plate the boundary conditions for all kinematic variables are of the same homogeneous Dirichlet type. Therefore, we will look for the solution in the function space $\mathcal{H}(P)$ defined as

$$\mathcal{H} = \mathbf{H}_0^1(P)^9. \qquad (6.19)$$

The space $\mathcal{H}$ is equipped with the inner product $\langle u, v \rangle_{\mathcal{H}}$:

$$\langle u, v \rangle_{\mathcal{H}} = \sum_{i=1}^{9} \langle u_i, v_i \rangle_{\mathbf{H}_0^1} \text{ for } u, v \in \mathcal{H}.$$

Relative to the metric

$$d(u, v) = \|u - v\|_{\mathcal{H}} \text{ for } u, v \in \mathcal{H},$$

induced by the norm $\|x\| = \sqrt{\langle x, x \rangle_{\mathcal{H}}}$, the space $\mathcal{H}$ is a complete metric space and, therefore, is a Hilbert space [20].

6.5 WEAK FORMULATION OF THE COSSERAT PLATE BENDING PROBLEM

In this section, we will derive the weak formulation of the clamped Cosserat plate bending problem. Let the system of the field equations (6.1) be written as

$$Lu = f(\eta).$$

Let us consider the dot product of both sides of the system and an arbitrary function $v \in \mathcal{H}$:

$$v \cdot Lu = v \cdot f(\eta),$$

and then integrate both sides of the obtained scalar equation over the plate P:

$$\int_P (v \cdot Lu)\,ds = \int_P (v \cdot f(\eta))\,ds.$$

Let us introduce a bilinear form $a(u,v) : \mathcal{H} \times \mathcal{H} \to \mathbb{R}$ and a linear form $b_{(\eta)}(v) : \mathcal{H} \to \mathbb{R}$ defined as

$$a(u,v) = \int_P (v \cdot Lu)\,ds, \tag{6.20}$$

$$b_{(\eta)}(v) = \int_P (v \cdot f(\eta))\,ds.$$

The expression

$$a(v,u) = \int_P \left(v_i L_{ij} u_j \right) ds,$$

is a summation over the terms of the form

$$a^{ij}(v_m, u_n) = \int_P \left(v_m \hat{L} u_n \right) ds,$$

where $v_m \in \mathcal{H}_m$, $u_n \in \mathcal{H}_n$ and $\hat{L}$ is a scalar differential operator.

There are 3 types of linear operators present in the field equations (6.5): operators of order zero, one and two, which are constant multiples of the following differential operators:

$$L^{(0)} = 1, \tag{6.21}$$

$$L^{(1)} = \frac{\partial}{\partial x_\alpha}, \tag{6.22}$$

$$L^{(2)} = -\nabla \cdot A\nabla. \tag{6.23}$$

These operators act on the components of the vector u and are multiplied by the components of the vector v. The obtained expressions are then integrated over the plate P:

$$\int_P \left(v_m L^{(0)} u_n\right) ds = \int_P (v_m u_n)\, ds, \tag{6.24}$$

$$\int_P \left(v_m L^{(1)} u_n\right) ds = \int_P \left(v_m \frac{\partial u_n}{\partial x_\alpha}\right) ds, \tag{6.25}$$

$$\int_P \left(v_m L^{(2)} u_n\right) ds = -\int_P (v_m (\nabla \cdot A\nabla) u_n)\, ds,$$

where $v_m \in \mathcal{H}_m$ and $u_n \in \mathcal{H}_n$.

The weak form of the second order operator is obtained by performing the corresponding integration by parts and taking into account that the test functions v_m vanish on the boundary ∂P:

$$\int_P \left(v_m L^{(2)} u_n\right) ds = -\int_P (v_m (\nabla \cdot A\nabla u_n))\, ds$$

$$= -\int_{\partial P} (A\nabla u_n \cdot n)\, v_m d\tau + \int_P (A\nabla u_n \cdot \nabla v_m)\, ds$$

$$= \int_P (A\nabla u_n \cdot \nabla v_m)\, ds. \tag{6.26}$$

The expression

$$b_{(\eta)}(v) = \int_P v_i f_i(\eta)\, ds,$$

represents a summation over the terms of the form:

$$\int_P v_m \hat{f}(\eta)\, ds.$$

Taking into account that the optimal solution of the field equations (6.5) minimizes the stress plate energy (3.42), the weak formulation for the clamped Cosserat plate bending problem consists of finding all $u \in \mathcal{H}$ and $\eta \in \mathbb{R}$ such that for all $v \in \mathcal{H}$

$$a(v, u) = b_{(\eta)}(v), \tag{6.27}$$

and that at the same time minimize the stress energy $\Phi(\mathcal{S}, \eta)$ given as (3.42).

> **Weak Formulation of the Clamped Cosserat Plate**
>
> Find all $u \in \mathcal{H}$ and $\eta \in \mathbb{R}$ that minimize the stress plate energy $\Phi(\mathcal{S}, \eta)$, subject to
>
> $$a(v, u) = b_{(\eta)}(v)$$
>
> for all $v \in \mathcal{H}$.

6.6 CONSTRUCTION OF THE FINITE ELEMENT SPACES

In this section, we will construct the finite element space, i.e. the finite-dimensional subspace $\mathcal{H}_h$ of the space $\mathcal{H}$, where we will be looking for an approximate Finite Element solution of the weak formulation (6.27).

Let us assume that the boundary ∂P is a polygonal curve. Let us make a triangulation of the domain P by subdividing P into l non-overlapping triangles K_i with m vertices N_j:

$$P = \bigcup_{i=1}^{l} K_i = K_1 \cup K_2 \cup \ldots \cup K_l,$$

such that no vertex of the triangular element lies on the edge of another triangle (see Figure 6.1).

The mesh parameter h is the greatest diameter among the elements K_i:

$$h = \max_{i=\overline{1,l}} d(K_i),$$

which for the triangular elements corresponds to the length of the longest side of the triangle.

We now define the finite dimensional space $\hat{\mathcal{H}}_h$ as a space of all continuous functions that are linear on each element K_j and vanish on the boundary:

$$\hat{\mathcal{H}}_h = \mathcal{H}_i^h = \left\{ v : v \in C(P), v \text{ is linear on every } K_j, v = 0 \text{ on } \partial P \right\}.$$

By definition $\mathcal{H}_i^h \subset \mathcal{H}_i$, and the finite element space $\mathcal{H}_h$ is then defined as

$$\mathcal{H}_h = \hat{\mathcal{H}}_h^9. \tag{6.28}$$

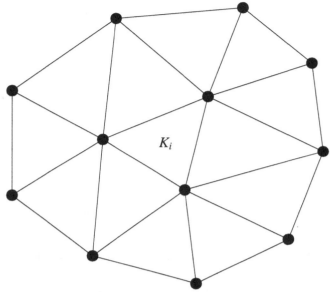

Figure 6.1 Example of the Finite Element triangulation of the domain P.

The approximate weak solution u^h can be found from the following Galerkin formulation of the clamped Cosserat plate bending problem that consists of finding all $u^h \in \mathcal{H}_h$ and $\eta \in \mathbb{R}$, such that for all $v^h \in \mathcal{H}_h$

$$a\left(v^h, u^h\right) = b_{(\eta)}\left(v^h\right),$$

and that at the same time minimize the stress plate energy $\Phi\left(\mathcal{S}, \eta\right)$ defined as (3.42).

Galerkin Formulation of the Clamped Cosserat Plate Bending Problem

Find all $u^h \in \mathcal{H}_h$ and $\eta \in \mathbb{R}$ that minimize the stress plate energy $\Phi\left(\mathcal{S}, \eta\right)$, subject to

$$a\left(v^h, u^h\right) = b_{(\eta)}\left(v^h\right) \tag{6.29}$$

for all $v^h \in \mathcal{H}_h$.

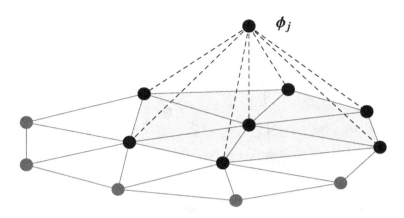

ϕ_j

Figure 6.2   Example of the Finite Element basis function ϕ_j.

The description of the function $v_i^h \in \mathcal{H}_i^h$ is provided by the values $v_i^h(N_k)$ at the nodes N_k ($k = \overline{1,m}$).

The set of basis functions $\{\phi_1, \phi_2, ..., \phi_m\}$ of each space $\mathcal{H}_i^h$ can be introduced as follows

$$\phi_j(N_k) = \delta_{jk}, \ j,k = \overline{1,m},$$

excluding the points N_k on the boundary ∂P. Therefore, we have that

$$\mathcal{H}_i^h = \text{span}\{\phi_1, \phi_2, ..., \phi_m\} = \left\{ v : v = \sum_{j=1}^{m} \alpha_j^{(i)} \phi_j \right\},$$

and the functions ϕ_j are nonzero only at the node N_j. The support of ϕ_j consists of all triangles K_i with the common node N_j (see Figure 6.2).

Since the spaces $\mathcal{H}_i^h$ are identical they will also have identical sets of basis functions ϕ_j ($j = \overline{1,m}$). Sometimes we will need to distinguish between the basis functions of different spaces assigning the superscript of the functions space to the basis function, i.e. the basis functions for the space $\mathcal{H}_i^h$ are ϕ_j^i. For computational purposes these superscripts will be dropped.

6.7 CALCULATION OF THE STIFFNESS MATRIX AND THE LOAD VECTOR

The bilinear form of the Galerkin formulation (6.6) is given as

$$a\left(v^h, u^h\right) = a^{ij}\left(v_i^h, u_j^h\right) = \int_P v_i^h L_{ij} u_j^h ds. \tag{6.30}$$

Since $u_j^h \in \mathcal{H}_j^h$, then there exist such constants $\alpha_p^{(j)} \in \mathbb{R}$ that

$$u_j^h = \alpha_p^{(j)} \phi_p^{(i)}.$$

Since the equation (6.30) is satisfied for all $v_i^h \in \mathcal{H}_i^h$, then it is also satisfied for all basis functions $\phi_k^{(i)}$ ($k = \overline{1,m}$):

$$a^{ij}\left(v_i^h, u_j^h\right) = a^{ij}\left(\phi_k^{(i)}, \alpha_p^{(j)} \phi_p^{(j)}\right) = \alpha_p^{(j)} a^{ij}\left(\phi_k^{(i)}, \phi_p^{(j)}\right),$$

where

$$a^{ij}(v, u) = \int_P v L_{ij} u \, ds. \tag{6.31}$$

Following [53], we define the block stiffness matrices K^{ij} ($i, j = \overline{1,9}$):

$$K^{ij} = \begin{bmatrix} a^{ij}\left(\phi_1^{(i)}, \phi_1^{(j)}\right) & \cdots & a^{ij}\left(\phi_1^{(i)}, \phi_m^{(j)}\right) \\ \vdots & \ddots & \vdots \\ a^{ij}\left(\phi_m^{(i)}, \phi_1^{(j)}\right) & \cdots & a^{ij}\left(\phi_m^{(i)}, \phi_m^{(j)}\right) \end{bmatrix}.$$

For computational purposes the superscripts of the basis functions can be dropped and the block stiffness matrices K^{ij} can be calculated as follows

$$K^{ij} = \begin{bmatrix} a^{ij}(\phi_1, \phi_1) & \cdots & a^{ij}(\phi_1, \phi_m) \\ \vdots & \ddots & \vdots \\ a^{ij}(\phi_m, \phi_1) & \cdots & a^{ij}(\phi_m, \phi_m) \end{bmatrix}.$$

Following [53], we define the block load vectors $F^i(\eta)$ ($i = \overline{1,9}$):

$$F^i(\eta) = \begin{bmatrix} b_{(\eta)}^i(\phi_1) \\ \vdots \\ b_{(\eta)}^i(\phi_m) \end{bmatrix},$$

and the solution block vectors α^i corresponding to the variable u_i^h $(i = \overline{1,9})$:

$$\alpha^i = \begin{bmatrix} \alpha_1^i \\ \vdots \\ \alpha_m^i \end{bmatrix}.$$

The equation (6.6) of the Galerkin formulation can be rewritten as

$$\left(K^{ij}\right)\alpha^i = F^j(\eta). \tag{6.32}$$

The global stiffness matrix consists of 81 block stiffness matrices K^{ij}, while the global load vector consists of 9 block load vectors $F^i(\eta)$ and the global displacement vector is represented by the 9 blocks of coefficients α^i. The entries of the block matrices K^{ij} and the block vectors $F^i(\eta)$ can be calculated as

$$K_{mn}^{ij} = \int_P \phi_m L_{ij} \phi_n \, ds,$$

$$F_m^i(\eta) = \int_P \phi_m f_i(\eta) \, ds.$$

The block matrix form of the equation (6.6) is given as

$$\begin{bmatrix} K^{11} & K^{12} & \cdots & K^{19} \\ K^{21} & K^{22} & \cdots & K^{29} \\ \vdots & \vdots & \ddots & \vdots \\ K^{91} & K^{92} & \cdots & K^{99} \end{bmatrix} \begin{bmatrix} \alpha^1 \\ \alpha^2 \\ \vdots \\ \alpha^9 \end{bmatrix} = \begin{bmatrix} F^1(\eta) \\ F^2(\eta) \\ \vdots \\ F^9(\eta) \end{bmatrix}$$

6.8 NUMERICAL RESULTS

We solve the parametric system field equations (6.1) for the clamped Cosserat plate with the boundary conditions (6.2)–(6.10) made of the polyurethane foam with the following values of the Lamé and Cosserat Elasticity parameters:

$$\lambda = 762.616,$$
$$\mu = 103.993,$$
$$\alpha = 4.333,$$
$$\beta = 39.975,$$
$$\gamma = 39.975,$$
$$\epsilon = 4.505.$$

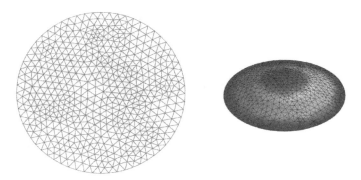

Figure 6.3 Circular clamped Cosserat plate of radius $R = 1.0$ m and thickness $h = 0.1$ m made of polyurethane foam under the sinusoidal load: the initial mesh and the isometric view of the resulting vertical deflection of the plate.

Figures 6.3–6.8 provide the visualization of the initial mesh and the Finite Element computation of the vertical deflection u_3 for the Cosserat plates of different shapes under different loads.

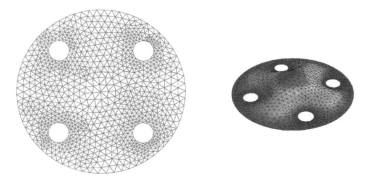

Figure 6.4 Circular clamped perforated Cosserat plate of radius $R = 1.0$ m and thickness $h = 0.1$ m made of polyurethane foam under the sinusoidal load: the initial mesh and the isometric view of the resulting vertical deflection of the plate.

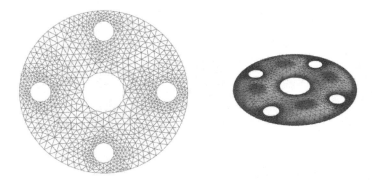

Figure 6.5 Circular clamped perforated Cosserat plate of radius $R = 1.0$ m and thickness $h = 0.1$ m made of polyurethane foam under the uniform load: the initial mesh and the isometric view of the resulting vertical deflection of the plate.

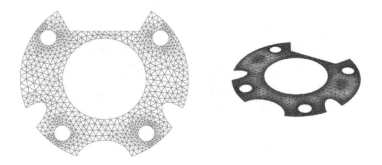

Figure 6.6 Clamped Cosserat polyurethane gasket under the uniform load: the initial mesh and the isometric view of the resulting vertical deflection of the plate.

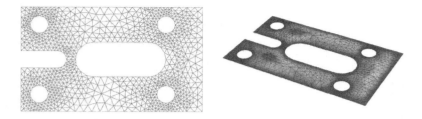

Figure 6.7 Clamped perforated Cosserat plate of size 10.0 m × 6.0 m × 0.1 m made of polyurethane foam under the uniform load: the initial mesh and the isometric view of the resulting vertical deflection of the plate.

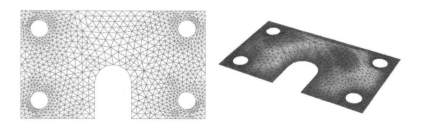

Figure 6.8 Clamped perforated Cosserat plate of size 10.0 m × 6.0 m × 0.1 m made of polyurethane foam under the uniform load: the initial mesh and the isometric view of the resulting vertical deflection of the plate.

CHAPTER 7

Validation of the Finite Element Computation

7.1   SCOPE OF THE CHAPTER

This chapter provides the numerical validation of the Finite Element Method for Cosserat Plates. We estimate the order of convergence of the main kinematic variables and validate the FEM computation for the homogeneous Dirichlet and the mixed Dirichlet-Neumann boundary conditions. We also present the validation of the Finite Element Method with the minimization procedure for the splitting parameter, for the problem of bending of the simply supported Cosserat plates. We show that for the piecewise linear approximation the convergence of the FEM achieves the optimal rates.

7.2   SAMPLE COSSERAT PLATE

Let us consider the Cosserat plate P to be a square plate $[0, a] \times [0, a]$ shown on Figure 7.1, with the boundary $G = G_1 \cup G_2 \cup G_3 \cup G_4$,

DOI: 10.1201/9781003190264-7

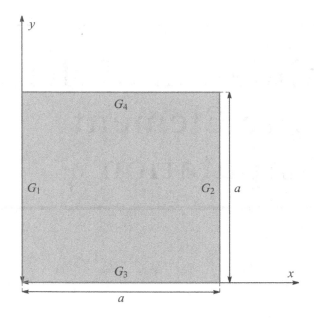

Figure 7.1 Sample Cosserat Plate.

where

$$G_1 = \{(x_1, x_2) : x_1 = 0, x_2 \in [0, a]\},$$
$$G_2 = \{(x_1, x_2) : x_1 = a, x_2 \in [0, a]\},$$
$$G_3 = \{(x_1, x_2) : x_1 \in [0, a], x_2 = 0\},$$
$$G_4 = \{(x_1, x_2) : x_1 \in [0, a], x_2 = a\},$$

The plate is assumed to be made of the polyurethane foam with the following values of the Lamé and Cosserat Elasticity parameters:

$$\lambda = 762.616,$$
$$\mu = 103.993,$$
$$\alpha = 4.333,$$
$$\beta = 39.975,$$
$$\gamma = 39.975,$$
$$\epsilon = 4.505.$$

We will solve the system of Cosserat plate bending field equations (3.93) written in the following form

$$LU = f(\eta), \tag{7.1}$$

where L is an elliptic differential operator defined in Chapter 2 that acts on the vector of nine kinematic variables

$$U = \left[\Psi_1, \Psi_2, W^*, \hat{W}, \Omega_1^*, \Omega_2^*, \hat{\Omega}_1, \hat{\Omega}_2, \Omega_3 \right],$$

and $f(\eta)$ is the right-hand side vector, that in general depends on the splitting parameter η.

For the case of the piecewise linear polynomials the optimal rate of convergence of the classical Finite Element Method in $\mathbf{H}^1$ norm is linear and in L_2 norm is quadratic. We will use the uniform refinement to create the sequence of triangulations and estimate the order of the error of the approximation of the FEM in $\mathbf{H}^1$ and L_2 norms.

7.3 VALIDATION OF THE FEM FOR DIRICHLET BOUNDARY CONDITIONS

Let us consider the homogeneous Dirichlet boundary conditions given in Table 7.1. We will assume the components of the solution U of the form:

$$\Psi_1 = A_1 \cdot \sin\left(\frac{\pi x_1}{a}\right) \cdot \sin\left(\frac{\pi x_2}{a}\right), \tag{7.2}$$

$$\Psi_2 = A_2 \cdot \sin\left(\frac{\pi x_1}{a}\right) \cdot \sin\left(\frac{\pi x_2}{a}\right), \tag{7.3}$$

$$W^* = A_3 \cdot \sin\left(\frac{\pi x_1}{a}\right) \cdot \sin\left(\frac{\pi x_2}{a}\right), \tag{7.4}$$

$$\hat{W} = A_4 \cdot \sin\left(\frac{\pi x_1}{a}\right) \cdot \sin\left(\frac{\pi x_2}{a}\right), \tag{7.5}$$

$$\Omega_1^* = A_5 \cdot \sin\left(\frac{\pi x_1}{a}\right) \cdot \sin\left(\frac{\pi x_2}{a}\right), \tag{7.6}$$

$$\Omega_2^* = A_6 \cdot \sin\left(\frac{\pi x_1}{a}\right) \cdot \sin\left(\frac{\pi x_2}{a}\right), \tag{7.7}$$

$$\hat{\Omega}_1 = A_7 \cdot \sin\left(\frac{\pi x_1}{a}\right) \cdot \sin\left(\frac{\pi x_2}{a}\right), \tag{7.8}$$

TABLE 7.1   Dirichlet boundary conditions for the square Cosserat plate.

Boundary $G_1 \cup G_2$	Boundary $G_3 \cup G_4$
$\Psi_1 = 0$	$\Psi_1 = 0$
$\Psi_2 = 0$	$\Psi_2 = 0$
$W^* = 0$	$W^* = 0$
$\hat{W} = 0$	$\hat{W} = 0$
$\Omega_1^* = 0$	$\Omega_1^* = 0$
$\Omega_2^* = 0$	$\Omega_2^* = 0$
$\hat{\Omega}_1 = 0$	$\hat{\Omega}_1 = 0$
$\hat{\Omega}_2 = 0$	$\hat{\Omega}_2 = 0$
$\Omega_3 = 0$	$\Omega_3 = 0$

$$\hat{\Omega}_2 = A_8 \cdot \sin\left(\frac{\pi x_1}{a}\right) \cdot \sin\left(\frac{\pi x_2}{a}\right), \tag{7.9}$$

$$\Omega_3 = A_9 \cdot \sin\left(\frac{\pi x_1}{a}\right) \cdot \sin\left(\frac{\pi x_2}{a}\right), \tag{7.10}$$

where A_i are constant $(i = 1, ..., 9)$.

Notice that the kinematic variables automatically satisfy the homogeneous Dirichlet boundary conditions given in Table 7.1. If we apply the elliptic differential operator L to the kinematic variables in the form (7.2)–(7.10), we can find the corresponding right-hand side function f and consider it as given in our validation. In this case, no minimization procedure is needed for the splitting parameter η.

The results of the error estimation of the Finite Element approximation in $\mathbf{H}^1$ and L_2 norms performed for the elastic parameters corresponding to the polyurethane foam are given in Tables 7.2 and 7.3 respectively. The convergence of the FEM achieves the optimal linear rate in $\mathbf{H}^1$ norm and the optimal quadratic rate in L_2 norm.

TABLE 7.2 The order of convergence in H^1 norm for the homogeneous Dirichlet boundary conditions.

Iteration Number	Number of Nodes	Diameter	Error in H^1 norm	Convergence Rate
0	177	0.302456	1.620369	
1	663	0.151228	0.711098	1.19
2	2565	0.075614	0.322016	1.14
3	10089	0.037807	0.150149	1.10
4	40017	0.018903	0.073481	1.03
5	159393	0.009451	0.036512	1.01

TABLE 7.3 The order of convergence in L_2 norm for the homogeneous Dirichlet boundary conditions.

Iteration Number	Number of Nodes	Diameter	Error in L_2 norm	Convergence Rate
0	177	0.302456	0.279484	
1	663	0.151228	0.069632	2.00
2	2565	0.075614	0.018175	1.94
3	10089	0.037807	0.004598	1.98
4	40017	0.018903	0.001153	2.00
5	159393	0.009451	0.000288	2.00

7.4 VALIDATION OF THE FEM FOR MIXED DIRICHLET-NEUMANN BOUNDARY CONDITIONS

Let us consider the mixed Dirichlet-Neumann boundary conditions given in Table 7.4. We will assume the components of the solution $\mathcal{U}$ of the form:

$$\Psi_1 = B_1 \cdot \cos\left(\frac{\pi x_1}{a}\right) \cdot \sin\left(\frac{\pi x_2}{a}\right), \tag{7.11}$$

$$\Psi_2 = B_2 \cdot \sin\left(\frac{\pi x_1}{a}\right) \cdot \cos\left(\frac{\pi x_2}{a}\right), \tag{7.12}$$

$$W^* = B_3 \cdot \sin\left(\frac{\pi x_1}{a}\right) \cdot \sin\left(\frac{\pi x_2}{a}\right), \tag{7.13}$$

$$\hat{W} = B_4 \cdot \sin\left(\frac{\pi x_1}{a}\right) \cdot \sin\left(\frac{\pi x_2}{a}\right), \tag{7.14}$$

$$\Omega_1^* = B_5 \cdot \sin\left(\frac{\pi x_1}{a}\right) \cdot \cos\left(\frac{\pi x_2}{a}\right), \tag{7.15}$$

$$\Omega_2^* = B_6 \cdot \cos\left(\frac{\pi x_1}{a}\right) \cdot \sin\left(\frac{\pi x_2}{a}\right), \tag{7.16}$$

$$\hat{\Omega}_1 = B_7 \cdot \sin\left(\frac{\pi x_1}{a}\right) \cdot \cos\left(\frac{\pi x_2}{a}\right), \tag{7.17}$$

$$\hat{\Omega}_2 = B_8 \cdot \cos\left(\frac{\pi x_1}{a}\right) \cdot \sin\left(\frac{\pi x_2}{a}\right), \tag{7.18}$$

$$\Omega_3 = B_9 \cdot \cos\left(\frac{\pi x_1}{a}\right) \cdot \cos\left(\frac{\pi x_2}{a}\right), \tag{7.19}$$

where B_i $(i = 1, ..., 9)$ are constant.

Notice that the kinematic variables automatically satisfy the mixed Dirichlet-Neumann boundary conditions given in Table 7.4. If we apply the elliptic differential operator L to the kinematic variables in the form (7.11)–(7.19), we can find the corresponding right-hand side function f and consider it as given in our validation. In this case, no minimization procedure is needed for the parameter η.

The results of the error estimation of the Finite Element approximation in $\mathbf{H}^1$ and L_2 norms performed for the elastic parameters corresponding to the polyurethane foam are given in Tables 7.5 and 7.6 respectively. The convergence of the FEM achieves the optimal linear rate in $\mathbf{H}^1$ norm and the optimal quadratic rate in L_2 norm.

TABLE 7.4 Mixed Dirichlet-Neumann boundary conditions for the square Cosserat plate.

Boundary $G_1 \cup G_2$	Boundary $G_3 \cup G_4$
$\frac{\partial \Psi_1}{\partial n} = 0$	$\frac{\partial \Psi_2}{\partial n} = 0$
$\Psi_2 = 0$	$\Psi_1 = 0$
$W^* = 0$	$W^* = 0$
$\hat{W} = 0$	$\hat{W} = 0$
$\Omega_1^* = 0$	$\Omega_2^* = 0$
$\frac{\partial \Omega_2^*}{\partial n} = 0$	$\frac{\partial \Omega_1^*}{\partial n} = 0$
$\hat{\Omega}_1 = 0$	$\hat{\Omega}_2 = 0$
$\frac{\partial \hat{\Omega}_2}{\partial n} = 0$	$\frac{\partial \hat{\Omega}_1}{\partial n} = 0$
$\Omega_3 = 0$	$\Omega_3 = 0$

TABLE 7.5 The order of Convergence in H^1 norm for the mixed Dirichlet-Neumann boundary conditions.

Iteration Number	Number of Nodes	Diameter	Error in H^1 norm	Convergence Rate
0	177	0.302456	0.236791	
1	663	0.151228	0.115809	**1.03**
2	2565	0.075614	0.054195	**1.09**
3	10089	0.037807	0.026233	**1.05**
4	40017	0.018903	0.012986	**1.01**
5	159393	0.009451	0.006475	**1.00**

TABLE 7.6 The order of Convergence in L_2 norm for the mixed Dirichlet-Neumann boundary conditions.

Iteration Number	Number of Nodes	Diameter	Error in L_2 norm	Convergence Rate
0	177	0.302456	6.214×10^{-2}	
1	663	0.151228	1.638×10^{-2}	1.92
2	2565	0.075614	4.219×10^{-3}	1.96
3	10089	0.037807	1.065×10^{-3}	1.99
4	40017	0.018903	2.678×10^{-4}	1.99
5	159393	0.009451	6.772×10^{-5}	1.98

7.5 VALIDATION OF THE FEM FOR SIMPLY SUPPORTED COSSERAT PLATE

Let us consider the simply supported boundary conditions given in Table 7.7. In Chapter 4, the components of the solution $\mathcal{U}$ were shown to be in the form:

$$\Psi_1 = C_1 \cdot \cos\left(\frac{\pi x_1}{a}\right) \cdot \sin\left(\frac{\pi x_2}{a}\right), \tag{7.20}$$

$$\Psi_2 = C_2 \cdot \sin\left(\frac{\pi x_1}{a}\right) \cdot \cos\left(\frac{\pi x_2}{a}\right), \tag{7.21}$$

$$W^* = C_3 \cdot \sin\left(\frac{\pi x_1}{a}\right) \cdot \sin\left(\frac{\pi x_2}{a}\right), \tag{7.22}$$

$$\hat{W} = C_4 \cdot \sin\left(\frac{\pi x_1}{a}\right) \cdot \sin\left(\frac{\pi x_2}{a}\right), \tag{7.23}$$

$$\Omega_1^* = C_5 \cdot \sin\left(\frac{\pi x_1}{a}\right) \cdot \cos\left(\frac{\pi x_2}{a}\right), \tag{7.24}$$

$$\Omega_2^* = C_6 \cdot \cos\left(\frac{\pi x_1}{a}\right) \cdot \sin\left(\frac{\pi x_2}{a}\right), \tag{7.25}$$

$$\hat{\Omega}_1 = C_7 \cdot \sin\left(\frac{\pi x_1}{a}\right) \cdot \cos\left(\frac{\pi x_2}{a}\right), \tag{7.26}$$

TABLE 7.7 Simply supported boundary conditions for the square Cosserat plate.

Boundary $G_1 \cup G_2$	Boundary $G_3 \cup G_4$
$\frac{\partial \Psi_1}{\partial n} = 0$	$\frac{\partial \Psi_2}{\partial n} = 0$
$\Psi_2 = 0$	$\Psi_1 = 0$
$W^* = 0$	$W^* = 0$
$\hat{W} = 0$	$\hat{W} = 0$
$\Omega_1^* = 0$	$\Omega_2^* = 0$
$\frac{\partial \Omega_2^*}{\partial n} = 0$	$\frac{\partial \Omega_1^*}{\partial n} = 0$
$\hat{\Omega}_1 = 0$	$\hat{\Omega}_2 = 0$
$\frac{\partial \hat{\Omega}_2}{\partial n} = 0$	$\frac{\partial \hat{\Omega}_1}{\partial n} = 0$
$\Omega_3 = 0$	$\Omega_3 = 0$

$$\hat{\Omega}_2 = C_8 \cdot \cos\left(\frac{\pi x_1}{a}\right) \cdot \sin\left(\frac{\pi x_2}{a}\right), \qquad (7.27)$$

$$\Omega_3 = C_9 \cdot \cos\left(\frac{\pi x_1}{a}\right) \cdot \cos\left(\frac{\pi x_2}{a}\right), \qquad (7.28)$$

where C_i ($i = 1,...9$) are constant.

The initial distribution of the pressure, as in Chapter 4, is assumed to be in the following sinusoidal form

$$p(x_1, x_2) = \sin\left(\frac{\pi x_1}{a}\right) \cdot \sin\left(\frac{\pi x_2}{a}\right).$$

We solve the system of field equations (7.1) using the Finite Element Algorithm for the Cosserat Plates discussed in Chapter 6. The algorithm includes the solution of the mixed Dirichlet-Neumann problem, calculations of the two-dimensional stresses and strains and the minimization procedure for the splitting parameter η. We compare the obtained numerical solution with the analytical solution for the square plate derived in Chapter 5. The

TABLE 7.8 The order of convergence in H^1 norm for the simply supported plate.

Iteration Number	Number of Nodes	Diameter	Error in H^1 norm	Convergence Rate
0	177	0.302456	0.256965	
1	663	0.151228	0.119234	**1.11**
2	2565	0.075614	0.054701	**1.12**
3	10089	0.037807	0.026301	**1.05**
4	40017	0.018903	0.012994	**1.01**
5	159393	0.009451	0.006476	**1.00**

estimation of the error in H^1 norm shows that for the simply supported plate the order of convergence is optimal (linear) for the piecewise linear elements. The estimation of the error in L_2 norm shows that for the simply supported plate the order of convergence is optimal (quadratic) for the piecewise linear elements. The estimation of the error in H^1 and L_2 norms performed for the elastic parameters corresponding to the polyurethane foam are given in Tables 7.8 and 7.9 respectively.

The initial triangulation of the simply supported Cosserat plate and its Finite Element computation with 177 elements is given in Figure 7.2. The comparison of the extrema of the displacements u_i and the microrotations φ_i calculated using the Finite Element Method with 320 thousand elements and the analytical solution for the Cosserat Plate Theory is provided in Table 7.10. The minimization procedure provides the approximation of the optimal value of the splitting parameter with a relative error of 0.09%.

TABLE 7.9 The order of convergence in L_2 norm for the simply supported plate.

Iteration Number	Number of Nodes	Diameter	Error in L_2 norm	Convergence Rate
0	177	0.302456	8.253×10^{-2}	
1	663	0.151228	2.260×10^{-2}	**1.87**
2	2565	0.075614	5.860×10^{-3}	**1.95**
3	10089	0.037807	1.482×10^{-3}	**1.98**
4	40017	0.018903	3.720×10^{-4}	**1.99**
5	159393	0.009451	9.355×10^{-5}	**1.99**

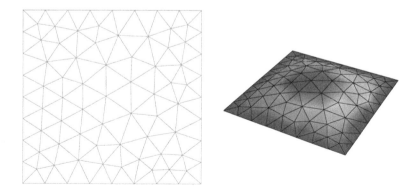

Figure 7.2 Finite Element computation of the vertical deflection for the simply supported Cosserat plate with 177 elements.

TABLE 7.10 The relative errors of the optimal value of the splitting parameter η and the maximum values of the displacements and microrotations calculated with 320 thousand elements.

	FEM Solution	Analytical Solution	Relative Error
Optimal η	0.040760	0.040799	**0.09%**
u_1	-0.014891	-0.014892	**0.03%**
u_2	-0.014891	-0.014892	**0.03%**
u_3	0.307641	0.307674	**0.04%**
φ_1	0.046767	0.046770	**0.03%**
φ_2	-0.046767	-0.046770	**0.03%**
φ_3	0.000000	0.000000	**0.00%**

CHAPTER 8

Unique Properties of Cosserat Plates

8.1   SCOPE OF THE CHAPTER

This chapter presents the unique properties of the Cosserat plate statics and dynamics. We show that these properties are the result of the presence of the microstructure in the plate. The numerical experiments show that the microstructure may increase the stiffness of the plates and change the influence of the plate size on its deformation. The microstructure is also responsible for the additional Cosserat resonances of the plate free vibration. Furthermore, the experiments show the dynamic anisotropy of Cosserat plates, i.e. the dependency of the Cosserat frequencies on the shape and the orientation of the microstructure elements.

8.2   COSSERAT SAMPLE PLATE

For the numerical experiments we will consider the low density rigid Cosserat plates made of polyurethane foam. The thickness of the plates is $h = 0.1$ m with the width to thickness ratio a/h varying from 5 to 30. The plates will have different levels of microstructure and different shapes of microelements. The polyurethane foam is assumed to have the density value

DOI: 10.1201/9781003190264-8

$\rho = 34 \text{ kg/m}^3$ and the Lamé and Cosserat Elasticity parameters:

$$\lambda = 762.616,$$
$$\mu = 103.993,$$
$$\alpha = 4.333,$$
$$\beta = 39.975,$$
$$\gamma = 39.975,$$
$$\epsilon = 4.505.$$

Let J_x, J_y, and J_z be the principal moments of microinertia of the microelements corresponding to the principal axes of their rotation. We assume that the quantities J_x, J_y, and J_z are constant throughout the plate. If the microelements are rotated around the z-axis by the angle θ, then the microinertia tensor $\mathbf{J}$ can be written as

$$\mathbf{J} = \begin{pmatrix} J_x \cos^2 \theta + J_y \sin^2 \theta & \left(J_x - J_y\right) \sin 2\theta & 0 \\ \left(J_x - J_y\right) \sin 2\theta & J_x \sin^2 \theta + J_y \cos^2 \theta & 0 \\ 0 & 0 & J_z \end{pmatrix}. \quad (8.1)$$

8.3 COSSERAT PLATE STATICS

The goal of the numerical experiments of the statics is to analyze the influence of the material microstructure on the bending of the Cosserat plates. In this section, we discuss the following properties of the Cosserat plates: plate stiffness and stress concentration.

8.3.1 Plate Stiffness

Stiffness of the plate reflects the amount of the plate's vertical deflection under an external force acting perpendicular to the plate's surface. We estimate the stiffness by evaluating the ratio

$$S_T = \frac{\max |\mathbf{F}|}{\max |u_3|}.$$

where $\mathbf{F}$ is the external force and u_3 is the vertical deflection.

In the numerical experiments the bending forces were chosen, such that the maxima of the forces and the resulting maxima of the vertical deflection take place in the geometrical centers of the

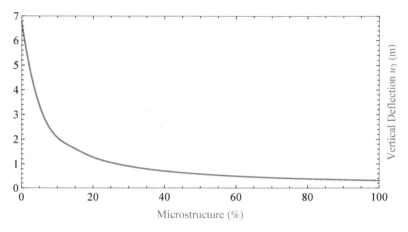

Figure 8.1   Influence of the microstructure on the vertical displacement u_3 (width-to-thickness ratio of the plate $a/h = 30$).

plates. We change the contribution of the Cosserat Elasticity parameters α, β, γ, ϵ and keep the Lamé parameters λ and μ constant. This allows us to evaluate the influence of the level of the microstructure on the plate bending.

Figure 8.1 confirms the important fact that if we consider the Cosserat plates with the same Lamé parameters, then the plates that have larger Cosserat Elasticity parameters will have higher stiffness. This phenomenon can be explained as follows. The Cosserat bending deformation includes rotations of the microelements. Thus, the elastic energy includes the energy of the microrotations. When the microstructure level increases the microrotations increase as well. Therefore, the contribution of the classic deformation to the total elastic energy is reduced and the plate deflection will become smaller. In turn, this implies that the stiffness of the plate increases.

We can interpret this result in a different way. Let us imagine that we run a blind numerical experiment with two plates: a classic plate and a Cosserat plate. We assume that both plates respond similarly to the bending force. If we treat the plate as a classic plate without microstructure, then we should call the corresponding Lamé parameters the pseudo-Lamé parameters. However, if we treat the plate as a Cosserat plate with microstructure, then the values of the elastic Lamé parameters will be less than the pseudo-Lamé parameters.

The numerical experiments also allow us to analyze the influence of the microstructure on the Cosserat plate bending. We gradually reduce the level of microstructure of the Cosserat plate by decreasing the values of the Cosserat Elasticity parameters. For example, 1% of the microstructure means 1% of the original values of the Cosserat Elasticity parameters. Figure 8.1 shows the influence of the microstructure on the vertical displacement u_3 for a square Cosserat plate with the width-to-thickness ratio $a/h = 30$.

Figure 8.2 shows how the scale of the microstructure in relation to the sample plate size influence the plate bending. The

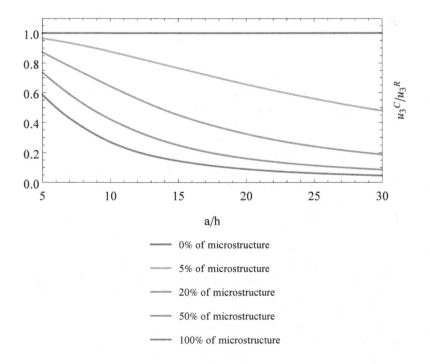

Figure 8.2 Microstructure size effect in bending: the ratio of the vertical displacement u_3^C of the Cosserat plate to the vertical displacement u_3^{MR} of the Mindlin-Reissner plate.

Cosserat plates of smaller thickness (or larger width) have smaller displacements than it would be expected based on the classic Mindlin-Reissner plate theory. The plates, therefore, become more stiff. This phenomenon is called the microstructure size effect.

8.3.2   Stress Concentration

Stress concentration occurs in the part of the plate, where the stress is significantly greater than in its surrounding area. We estimate the stress concentration factor by the following ratio

$$K = \frac{\sigma_{max}}{\sigma_{ref}}, \tag{8.2}$$

where σ_{max} is the highest stress and σ_{ref} is the reference stress of the cross section.

In our numerical experiments, we consider a square plate of size $a \times a$ and thickness h, with a hole located in the center of the plate as shown in Figure 8.3. The hole consists of a square of size $2r \times 2r$ and two semicircles of the radius r. The diameter of the hole is defined as $d = 4r$.

We use the Finite Element Method discussed in Chapter 5 to solve the problem of the Cosserat plate bending, evaluate the corresponding stress distributions and measure the stress concentration factors around the hole. Figure 8.4 shows the distribution of the stress σ_{11} and the stress concentration around the hole for the square polyurethane foam plate 10.0 cm × 10.0 cm × 1.0 cm.

For the Cosserat plates the stress around the hole is always smaller than the classical value. The comparison of the stress concentration factors of the Cosserat and Mindlin-Reissner plates is given in Figure 8.5. The values of the ratio of the hole diameter d to the plate width d is considered to be in the range from 0.1 to 0.5. The figure confirms that the small holes exhibit less stress concentration than the larger holes as it is expected on the basis of the classical theory [46] and results in the higher stiffness. This is consistent with the microstructure size effect for the vertical deflection of the Cosserat plates.

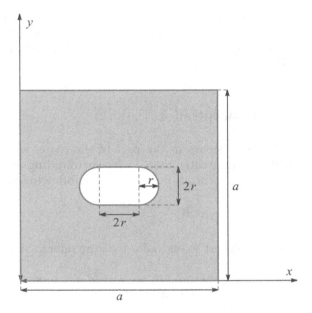

Figure 8.3 Square plate of size $a \times a$ and thickness h with a hole located in the center of the plate.

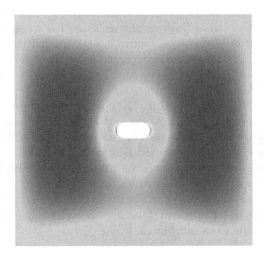

Figure 8.4 The distribution of the stress σ_{11} for the square polyurethane foam plate of size 10.0 cm × 10.0 cm × 1.0 cm demonstrates the stress concentration around the hole.

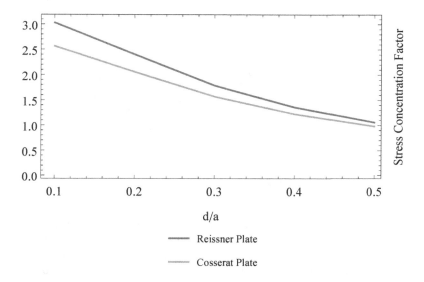

Figure 8.5   The distribution of the stress σ_{11} for the square polyurethane foam plate 10.0 cm × 10.0 cm × 1.0 cm demonstrates the stress concentration around the hole.

8.4   COSSERAT PLATE DYNAMICS

The goal of the numerical experiments of the dynamics is to analyze the influence of the material microstructure on the vibration of the Cosserat plates. In this section, we discuss the following properties of the Cosserat plates: plate resonances and dynamic anisotropy.

8.4.1   Plate Resonances

We will consider the microstructure consisting of the microelements shaped as balls or ellipsoids (see Figures 8.6–8.8).

The numerical experiments show that the Cosserat plates have four basic resonance frequencies: ω_K, ω_{MR}, ω_{C1}, and ω_{C2}. The first frequency ω_K is associated with the flexural motion and is an analog of the resonance of the Kirchhoff plate. For this reason we will call it the Kirchhoff resonance frequency. The second frequency ω_{MR} is associated with the rotation of the middle plane and is an analog of the resonance of the Mindlin-Reissner plate,

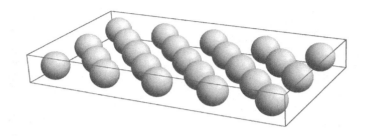

Figure 8.6   Ball-shaped microelements (not to scale drawing).

so we will call it the Mindlin-Reissner resonance frequency. The numerical experiments demonstrate that the microstructure may significantly affect the Kirchhoff frequency, while it has very little effect on the Mindlin-Reisner frequency. The next two frequencies ω_{C1} and ω_{C2} are associated directly with the microstructure. We will call these them the Cosserat frequencies. The experimental data confirm that the microstructure has a major effect on the Cosserat resonance frequencies.

The numerical experiments allow us to analyze the influence of the microstructure on the Cosserat plate vibration. We gradually reduce the level of the microstructure of the Cosserat plate by decreasing the values of Cosserat Elasticity parameters. For example, 1% of the microstructure means 1% of the original values of the Cosserat Elasticity parameters. Figure 8.9 shows the influence of the microstructure on the Kirchhoff resonance frequency

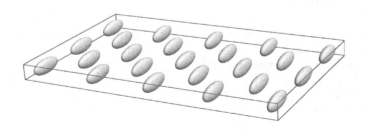

Figure 8.7   Ellipsoid microelements (not to scale drawing).

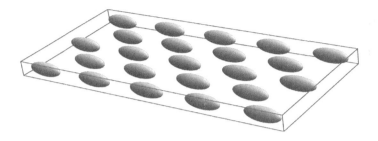

Figure 8.8   Ellipsoid microelements (not to scale drawing).

ω_K for a square Cosserat plate with the width-to-thickness ratio $a/h = 30$. Figure 8.10 shows the influence of the microstructure on the Mindlin-Reissner resonance frequency ω_{MR} for a square Cosserat plate with the width-to-thickness ratio $a/h = 30$.

Figure 8.11 shows how the scale of the microstructure in relation to the sample plate size influences the plate vibration. The Cosserat plates of smaller thickness (or larger width) have higher frequencies than it would be expected based on the classic Mindlin-Reissner theory. The plates, therefore, become more stiff. Similar experimental behavior was reported in [44] for torsion

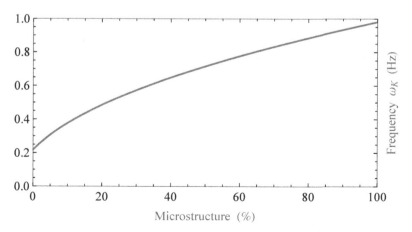

Figure 8.9   Effect of the microstructure on the Kirchhoff frequency ω_K (width-to-thickness ratio of the plate $a/h = 30$).

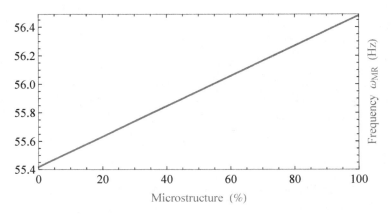

Figure 8.10 Effect of the microstructure on the Mindlin Reissner frequency ω_{MR} (width-to-thickness ratio of the plate $a/h = 30$).

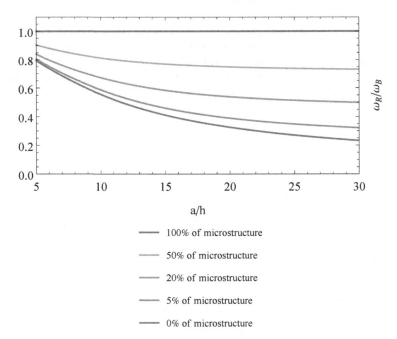

Figure 8.11 Microstructure size effect in vibration: the ratio of the Kirchhoff frequency ω_K to the base Kirchhoff frequency ω_B with 100% of microstructure present.

and bending of cylindrical Cosserat rods. This phenomenon is called the microstructure size effect. The size effect of the free vibration of the Cosserat plates is consistent with the size effect of the bending of Cosserat plates [41], [27].

8.4.2 Dynamic Anisotropy

In this section, we discuss the dynamic anisotropy of the Cosserat plates. We associate the anisotropy with the tensor J of the microelements microinertia. Table 8.1 contains the values of the principle moments of microinertia tensor J for different shapes of microelements. The values of the Kirchhoff frequency ω_K, Mindlin-Reissner frequency ω_{MR} and Cosserat frequencies ω_{C1} and ω_{C2} for different shapes of microelements are given in Table 8.2.

We perform the numerical experiments for different angles of orientation of the microelements. The principle moments of microinertia of the microelements are $J_x = 0.0001$, $J_y = 0.001$, $J_z = 0.001$, which represents the horizontally stretched ellipsoids E1. The corresponding frequencies are given in Table 8.3. In the case when the microelements are not aligned with the edges of the plate, the Cosserat Plate Theory predicts some additional

TABLE 8.1 Values of the principle moments of microinertia tensor J for different shapes of microelements.

Microelement Shape	J_x	J_y	J_z
Ball B1	0.001	0.001	0.001
Ball B2	0.0001	0.0001	0.0001
Vertical Ellipsoid E1	0.001	0.001	0.0001
Horizontal Ellipsoid E2	0.0001	0.001	0.001
Horizontal Ellipsoid E3	0.001	0.0001	0.001

TABLE 8.2 Values of the Kirchhoff frequency ω_K, Mindlin-Reissner frequency ω_{MR} and Cosserat frequencies ω_{C1} and ω_{C2} for different shapes of microelements.

Shape	Kirchhoff Frequency ω_K (Hz)	Mindlin-Reissner Frequency ω_{MR} (Hz)	Cosserat Frequency ω_{C1} (Hz)	Cosserat Frequency ω_{C2} (Hz)
B1	0.97999	56.5352	338.978	501.139
B2	0.97999	56.5352	1071.86	1584.72
E1	0.97999	56.5352	338.978	501.139
E2	0.97999	56.5352	394.102	1363.01
E3	0.97999	56.5352	394.102	1363.01

Cosserat frequencies related with the microstructure of the material.

The summary of the main properties of the Cosserat plates is provided in Table 8.4.

TABLE 8.3 Values of the Cosserat frequency ω_{C1} splitting into two frequencies ω_{C1A} and ω_{C1B} for different angles of microelement orientation θ.

Microelement Orientation Angle θ	Cosserat Frequency ω_{C1A} (Hz)	Cosserat Frequency ω_{C1B} (Hz)
0°	394.102	394.102
5°	369.404	417.164
10°	346.916	432.925
15°	328.015	438.614
20°	312.816	436.711
25°	301.018	431.321
30°	292.254	425.372
35°	286.216	420.436
40°	282.682	417.264
45°	281.518	416.177
50°	282.682	417.264
55°	286.216	420.436
60°	292.254	425.372
65°	301.018	431.321
70°	312.816	436.711
75°	328.015	438.614
80°	346.916	432.925
85°	369.404	417.164
90°	394.102	394.102

TABLE 8.4   Summary of the main properties of the Cosserat plates.

Microstructure Presence	Level of Microstructure	Shape of Microelements	Influence on Statics	Influence on Dynamics
Absent	0%	N/A	Classic Deformation (Mindlin-Reissner Plate)	Classic Vibration: only Kirchhoff and Mindlin-Reissner Frequencies Present
Present	100%	Fixed	Cosserat Deformation (Cosserat Plate)	Cosserat Vibration: Appear Cosserat Frequencies
Present	Increasing 0% to 100%	Fixed	Stiffness Increases; Stress Concentration Decreases	Kirchhoff, Mindlin-Reissner and Cosserat Frequencies Increase
Present	100%	Variable	No Influence	No influence on Kirchhoff and Mindlin-Reissner Frequencies; Influences only Cosserat Frequencies

Appendix A

IN PLATE TWISTING DEFORMATION OF COSSERAT PLATES

We consider the thin plate $P \subset \left(\Gamma \times \left[-\frac{h}{2}, \frac{h}{2}\right]\right) \cup T \cup B$, where h is the thickness of the plate. The sets T and B are the top and bottom surfaces contained in the planes $x_3 = h/2$, $x_3 = -h/2$ respectively, and the curve Γ is the plate boundary in the middle plane $x_3 = 0$.

The set of points $P = \left(\Gamma \times \left[-\frac{h}{2}, \frac{h}{2}\right]\right) \cup T \cup B$ forms the entire surface of the plate. We consider the boundary $\Gamma = \Gamma_u \cup \Gamma_\sigma$. The set of points $\Gamma_u \times \left[-\frac{h}{2}, \frac{h}{2}\right]$ represents the lateral part of the boundary where displacements and microrotations are prescribed. The set of points $\Gamma_\sigma \times \left[-\frac{h}{2}, \frac{h}{2}\right]$ represents the lateral part of the boundary where stress and couple stress are prescribed. We will also use the notation P_0 for the middle plane internal domain of the plate.

We will consider the plate under the pure twisting momentum at the top and bottom of the plate, i.e:

$$\sigma_{33}(x_1, x_2, h/2) = 0, \sigma_{3\beta}(x_1, x_2, h/2) = 0, \tag{A.1}$$

$$\sigma_{33}(x_1, x_2, -h/2) = 0, \sigma_{3\beta}(x_1, x_2, -h/2) = 0, \tag{A.2}$$

$$\mu_{33}(x_1, x_2, h/2) = \mu^t(x_1, x_2), \tag{A.3}$$

$$\mu_{33}(x_1, x_2, -h/2) = \mu^b(x_1, x_2), \tag{A.4}$$

where $(x_1, x_2) \in P_0$.

Our stress assumptions are given as

$$\sigma_{\alpha\beta} = n_{\alpha\beta}(x_1, x_2), \tag{A.5}$$

and couple stress assumptions

$$\mu_{\alpha 3} = m_\alpha^*(x_1, x_2), \tag{A.6}$$

$$\mu_{33} = v_1(x_1, x_2)\zeta + v_2(x_1, x_2), \tag{A.7}$$

where the functions $v_1(x_1, x_2)$ and $v_2(x_1, x_2)$ are defined by the bounsary conditions

$$v_1(x_1, x_2) = \frac{1}{2}\left(\mu^t(x_1, x_2) - \mu^b(x_1, x_2)\right), \tag{A.8}$$

$$v_2(x_1, x_2) = \frac{1}{2}\left(\mu^t(x_1, x_2) + \mu^b(x_1, x_2)\right). \tag{A.9}$$

The corresponding kinematic assumptions

$$u_\alpha = \hat{\psi}_\alpha(x_1, x_2), \tag{A.10}$$

$$\phi_3 = \hat{\omega}_3(x_1, x_2). \tag{A.11}$$

Now, the plate stress energy density by the formula

$$\Phi_T = \frac{h}{2}\int_{-1}^{1}\Phi_T\{\sigma, \mu\}d\zeta, \tag{A.12}$$

where

$$\Phi_T\{\sigma, \mu\} = \frac{\mu' + \alpha'}{2}\sigma_{\alpha\beta}\sigma_{\alpha\beta} + \frac{\mu' - \alpha'}{2}\sigma_{\alpha\beta}\sigma_{\beta\alpha} + \frac{\lambda'}{2}\sigma_{\alpha\alpha}\sigma_{\beta\beta}$$
$$+ \frac{\gamma' + \epsilon'}{2}\mu_{i3}\mu_{i3} + \frac{\gamma' - \epsilon'}{2}\mu_{i3}\mu_{3i} + \frac{\beta'}{2}\mu_{33}\mu_{33}. \tag{A.13}$$

Taking into account the stress and couple stress assumptions (B.7) and (B.8) and by the integrating $\Phi\{\sigma, \mu\}$ with respect ζ on $[-1, 1]$ we obtain the explicit plate stress energy density expression in the form:

$$\hat{\Phi}_T(\mathcal{S}_T) = -\frac{\lambda}{4h\mu(3\lambda + 2\mu)}(N_{\alpha\alpha})(N_{\beta\beta}) \tag{A.14}$$

$$+ \frac{\alpha}{4h\alpha\mu}(N_{\alpha\beta})^2 + \frac{\gamma + \epsilon}{24h^3\gamma\epsilon}(M_\alpha^*)^2$$

$$- \frac{\beta h}{4\gamma(3\beta + 2\gamma)}(v_1^2 + v_2^2)$$

$$+ \frac{h}{12\gamma}\left[(3v_1^2 + v_2^2)\right] \tag{A.15}$$

where the Cosserat stress set $\mathcal{S}_T$ is defined as

$$\mathcal{S}_T = \left[N_{\alpha\beta}, M_\alpha^*,\right], \tag{A.16}$$

where

$$N_{\alpha\beta} = \frac{h}{2} \int_{-1}^{1} \sigma_{\alpha\beta} d\zeta = h n_{\alpha\beta},$$

$$M_\alpha^* = \frac{h}{2} \int_{-1}^{1} \mu_{\alpha 3} d\zeta = h m_\alpha^*.$$

Here N_{11} and N_{22} are the bending forces, N_{12} and N_{21} the twisting forces, M_α^* the Cosserat shear couple-stress resultants.

Let us now evaluate the density of the work over the boundary $\Gamma_u \times [-h/2, h/2]$

$$\mathcal{W}_1 = \frac{h}{2} \int_{-1}^{1} [\boldsymbol{\sigma_n} \cdot \mathbf{u} + \boldsymbol{\mu_n} \cdot \boldsymbol{\phi}] d\zeta \qquad (A.17)$$

Taking into account the stress, couple stress assumptions, and kinematic assumptions we can rewrite $\mathcal{W}_1$ as follows

$$\mathcal{W}_1 = \mathcal{S}_T^n \cdot \mathcal{U}_T = \check{N}_\alpha \hat{\Psi}_\alpha + \check{M}^* \hat{\Omega}_3, \qquad (A.18)$$

where the sets $\mathcal{S}_T^n$ and $\mathcal{U}_T$ are defined as

$$\mathcal{S}_T^n = \left[\check{N}_\alpha, \check{M}^* \right],$$

$$\mathcal{U}_T = \left[\hat{\Psi}_\alpha, \hat{\Omega}_3 \right]$$

and

$$\check{N}_\alpha = N_{\alpha\beta} n_\beta,$$

$$\check{M}^* = M_\beta^* n_\beta,$$

In the above n_β is the outward unit normal vector to Γ_u, and

$$\hat{\Psi}_\alpha = \hat{\psi}_\alpha,$$

$$\hat{\Omega}_3 = \hat{\omega}_3.$$

Now we define the density of the work done by the stress and couple stress over the Cosserat strain field

$$\mathcal{W}_2 = \frac{h}{2} \int_{-1}^{1} (\boldsymbol{\sigma} \cdot \boldsymbol{\gamma} + \boldsymbol{\mu} \cdot \boldsymbol{\chi}) d\zeta. \qquad (A.19)$$

Substituting stress and couple stress assumptions and integrating the expression (B.25) we obtain

$$\mathcal{W}_2 = \mathcal{S}_T \cdot \mathcal{E}_T = N_{\alpha\beta} v_{\alpha\beta} + M_\alpha^* \tau_\alpha, \qquad (A.20)$$

where $\mathcal{E}_T$ is the Cosserat plate strain set of the the weighted averages of strain and torsion tensors

$$\mathcal{E}_T = \left[v_{\alpha\beta}, \tau_\alpha \right].$$

Here the components of $\mathcal{E}_T$ are

$$v_{\alpha\beta} = \frac{1}{2} \int_{-1}^{1} \gamma_{\alpha\beta} d\zeta_3,$$

$$\tau_\alpha = \frac{1}{2} \int_{-1}^{1} \chi_{3\alpha} d\zeta$$

The components of Cosserat plate strain can also be represented in terms of the components of set $\mathcal{U}$ by the following strain-displacement relation of the Cosserat plate

$$v_{\alpha\beta} = \hat{\Psi}_{\beta,\alpha} + \varepsilon_{3\alpha\beta} \hat{\Omega}_3,$$

$$\tau_\alpha = \hat{\Omega}_{3,\alpha}$$

Now, we are able to postulate the HPR variational principle for the twisting of Cosserat plate P.

Let $\mathcal{A}$ denote the set of all admissible states that satisfy the Cosserat plate strain-displacement relations and let Θ be the functional on $\mathcal{A}$ defined as

$$\Theta(s) = U_K^S$$
$$- \int_{P_0} (\mathcal{S}_T \cdot \mathcal{E}_T + v_1 \hat{\Omega}_3) da$$
$$+ \int_{\Gamma_\sigma} \mathcal{S}_T^o \cdot \left(\mathcal{U}_T - \mathcal{U}_T^o \right) ds$$
$$+ \int_{\Gamma_u} \mathcal{S}_T^n \cdot \mathcal{U}_T ds,$$

for every $s = [\mathcal{U}_T, \mathcal{E}_T, \mathcal{S}_T] \in \mathcal{A}$.
Then
$$\delta\Theta(s) = 0$$

is equivalent to the following plate twisting equilibrium system of equations

$$N_{\alpha\beta,\alpha} = 0, \tag{A.21}$$

$$M_{\alpha,\alpha}^* + \epsilon_{3\beta\gamma} N_{\beta\gamma} + v_1 = 0, \tag{A.22}$$

with the resultant traction boundary conditions at Γ_σ:

$$N_{\alpha\beta}n_\beta = N_{o\alpha},$$
$$M_\alpha^* n_\alpha = M_o^*,$$

and the resultant displacement boundary conditions

$$\mathcal{U}_T^o = \left[\hat{\Psi}_{o\alpha}, \hat{\Omega}_{o3}\right]$$

at Γ_u

$$\hat{\Psi}_\alpha = \hat{\Psi}_{o\alpha},$$
$$\hat{\Omega}_3 = \hat{\Omega}_{o3}.$$

and the constitutive formulas

$$\upsilon_{\alpha\alpha} = \frac{\partial\Phi}{\partial N_{\alpha\alpha}} = \frac{\lambda+\mu}{h\mu(3\lambda+2\mu)}N_{\alpha\alpha} - \frac{\lambda}{2h\mu(3\lambda+2\mu)}N_{\beta\beta},$$

$$\upsilon_{\alpha\beta} = \frac{\partial\Phi}{\partial N_{\alpha\beta}} = \frac{\alpha+\mu}{4h\alpha\mu}N_{\alpha\beta} + \frac{\alpha-\mu}{4h\alpha\mu}N_{\beta\alpha},$$

$$\tau_\alpha = \frac{\partial\Phi}{\partial M_\alpha^*} = \frac{\gamma+\epsilon}{4h\gamma\epsilon}M_\alpha^*.$$

Appendix B

IN PLATE TWISTING VIBRATION OF COSSERAT PLATES

We consider the thin plate $P \subset \left(\Gamma \times \left[-\frac{h}{2}, \frac{h}{2} \right] \right) \cup T \cup B$, where h is the thickness of the plate. The sets T and B are the top and bottom surfaces contained in the planes $x_3 = h/2$, $x_3 = -h/2$ respectively, and the curve Γ is the plate boundary in the middle plane $x_3 = 0$.

The set of points $P = \left(\Gamma \times \left[-\frac{h}{2}, \frac{h}{2} \right] \right) \cup T \cup B$ forms the entire surface of the plate. We consider the boundary $\Gamma = \Gamma_u \cup \Gamma_\sigma$. The set of points $\Gamma_u \times \left[-\frac{h}{2}, \frac{h}{2} \right]$ represents the lateral part of the boundary where displacements and microrotations are prescribed. The set of points $\Gamma_\sigma \times \left[-\frac{h}{2}, \frac{h}{2} \right]$ represents the lateral part of the boundary where stress and couple stress are prescribed. We will also use the notation P_0 for the middle plane internal domain of the plate.

We will consider the plate under the pure twisting momentum at the top and bottom of the plate, i.e:

$$\sigma_{33}\left(x_1, x_2, h/2, t\right) = 0, \tag{B.1}$$

$$\sigma_{3\beta}\left(x_1, x_2, h/2, t\right) = 0, \tag{B.2}$$

$$\sigma_{33}\left(x_1, x_2, -h/2, t\right) = 0, \tag{B.3}$$

$$\sigma_{3\beta}\left(x_1, x_2, -h/2, t\right) = 0, \tag{B.4}$$

$$\mu_{33}\left(x_1, x_2, h/2, t\right) = \mu^t\left(x_1, x_2, t\right), \tag{B.5}$$

$$\mu_{33}\left(x_1, x_2, -h/2, t\right) = \mu^b\left(x_1, x_2, t\right), \tag{B.6}$$

where $(x_1, x_2) \in P_0$.

Our stress assumptions are given as

$$\sigma_{\alpha\beta} = n_{\alpha\beta}\left(x_1, x_2, t\right), \tag{B.7}$$

and couple stress assumptions

$$\mu_{\alpha 3} = m_\alpha^*(x_1, x_2, t), \tag{B.8}$$

$$\mu_{33} = v_1\left(x_1, x_2, t\right)\zeta + v_2\left(x_1, x_2, t\right), \tag{B.9}$$

where the functions $v_1(x_1, x_2, t)$ and $v_2(x_1, x_2, t)$ are defined by the boundary conditions

$$v_1(x_1, x_2, t) = \frac{1}{2}\left(\mu^t(x_1, x_2, t) - \mu^b(x_1, x_2, t)\right), \tag{B.10}$$

$$v_2(x_1, x_2, t) = \frac{1}{2}\left(\mu^t(x_1, x_2, t) + \mu^b(x_1, x_2, t)\right). \tag{B.11}$$

The internal kinetic energy is

$$\int_{B_0} K_T\left\{\frac{\partial u}{\partial t}, \frac{\partial \phi}{\partial t}\right\} dV, \tag{B.12}$$

where

$$K_T\left\{\frac{\partial u}{\partial t}, \frac{\partial \phi}{\partial t}\right\} = \frac{\rho}{2}\left(\frac{\partial u}{\partial t}\right)^2 + \frac{J}{2}\left(\frac{\partial \phi}{\partial t}\right)\left(\frac{\partial \phi}{\partial t}\right) \tag{B.13}$$

Now, the plate stress energy given by the formula

$$\hat{K}_T = \frac{h}{2}\int_{-1}^{1} K_T\left\{\frac{\partial u}{\partial t}, \frac{\partial \phi}{\partial t}\right\} d\zeta, \tag{B.14}$$

and using the kinematic assumptions kinetic

$$u_\alpha = \hat{\psi}_\alpha(x_1, x_2, t), \tag{B.15}$$

$$\phi_3 = \hat{\omega}_3(x_1, x_2, t) \tag{B.16}$$

becomes

$$\hat{K}_T = \frac{\rho h}{2}\left(\frac{\partial \hat{\Psi}_1}{\partial t}\right)^2 + \frac{\rho h}{2}\left(\frac{\partial \hat{\Psi}_2}{\partial t}\right)^2 + \frac{J_{33}h}{2}\left(\frac{\partial \hat{\Omega}_3}{\partial t}\right)^2. \tag{B.17}$$

where

$$\hat{\Psi}_\alpha = \hat{\psi}_\alpha,$$

$$\hat{\Omega}_3 = \hat{\omega}_3.$$

Now, the plate stress energy by the formula

$$\hat{\Phi}_T = \frac{h}{2}\int_{-1}^{1} \Phi_T\{\sigma, \mu\} d\zeta, \tag{B.18}$$

where

$$\Phi_T\{\sigma, \mu\} = \frac{\mu' + \alpha'}{2}\sigma_{\alpha\beta}\sigma_{\alpha\beta} + \frac{\mu' - \alpha'}{2}\sigma_{\alpha\beta}\sigma_{\beta\alpha} + \frac{\lambda'}{2}\sigma_{\alpha\alpha}\sigma_{\beta\beta}$$
$$+ \frac{\gamma' + \epsilon'}{2}\mu_{i3}\mu_{i3} + \frac{\gamma' - \epsilon'}{2}\mu_{i3}\mu_{3i} + \frac{\beta'}{2}\mu_{33}\mu_{33}. \tag{B.19}$$

Taking into account the stress and couple stress assumptions (B.7) and (B.8) and by the integrating $\Phi\{\sigma,\mu\}$ with respect ζ on $[-1,1]$ we obtain the explicit plate stress energy density expression in the form:

$$\hat{\Phi}_T(\mathcal{S}_T) = -\frac{\lambda}{4h\mu(3\lambda + 2\mu)}(N_{\alpha\alpha})(N_{\beta\beta}) \qquad (B.20)$$

$$+ \frac{\alpha}{4h\alpha\mu}(N_{\alpha\beta})^2 + \frac{\gamma + \epsilon}{24h^3\gamma\epsilon}(M_\alpha^*)^2$$

$$- \frac{\beta h}{4\gamma(3\beta + 2\gamma)}(v_1^2 + v_2^2)$$

$$+ \frac{h}{12\gamma}\left[(3v_1^2 + v_2^2)\right] \qquad (B.21)$$

where the Cosserat stress set $\mathcal{S}_T$ is defined as

$$\mathcal{S}_T = \left[N_{\alpha\beta}, M_\alpha^*,\right], \qquad (B.22)$$

where

$$N_{\alpha\beta} = \frac{h}{2}\int_{-1}^{1}\sigma_{\alpha\beta}d\zeta = hn_{\alpha\beta},$$

$$M_\alpha^* = \frac{h}{2}\int_{-1}^{1}\mu_{\alpha 3}d\zeta = hm_\alpha^*.$$

Here N_{11} and N_{22} are the bending forces, N_{12} and N_{21} the twisting forces, M_α^* the Cosserat shear couple-stress resultants.

Let us now evaluate the density of the work over the boundary $\Gamma_u \times [-h/2, h/2]$

$$\mathcal{W}_1 = \frac{h}{2}\int_{-1}^{1}[\sigma_\mathbf{n} \cdot \mathbf{u} + \mu_\mathbf{n} \cdot \phi]d\zeta \qquad (B.23)$$

Taking into account the stress, couple stress assumptions and kinematic assumptions we can rewrite $\mathcal{W}_1$ as follows

$$\mathcal{W}_1 = \mathcal{S}_T^n \cdot \mathcal{U}_T = \check{N}_\alpha\hat{\Psi}_\alpha + \check{M}^*\hat{\Omega}_3, \qquad (B.24)$$

where the sets $\mathcal{S}_T^n$ and $\mathcal{U}_T$ are defined as

$$\mathcal{S}_T^n = \left[\check{N}_\alpha, \check{M}^*\right],$$

$$\mathcal{U}_T = \left[\hat{\Psi}_\alpha, \hat{\Omega}_3\right]$$

and

$$\check{N}_\alpha = N_{\alpha\beta} n_\beta,$$
$$\check{M}^* = M_\beta^* n_\beta,$$

Now we define the density of the work done by the stress and couple stress over the Cosserat strain field

$$\mathcal{W}_2 = \frac{h}{2} \int_{-1}^{1} (\sigma \cdot \gamma + \mu \cdot \chi) \, d\zeta. \qquad (B.25)$$

Substituting stress and couple stress assumptions and integrating the expression (B.25) we obtain

$$\mathcal{W}_2 = \mathcal{S}_T \cdot \mathcal{E}_T = N_{\alpha\beta} v_{\alpha\beta} + M_\alpha^* \tau_\alpha, \qquad (B.26)$$

where $\mathcal{E}_T$ is the Cosserat plate strain set of the the weighted averages of strain and torsion tensors

$$\mathcal{E}_T = \left[v_{\alpha\beta}, \tau_\alpha \right].$$

Here the components of $\mathcal{E}_T$ are

$$v_{\alpha\beta} = \frac{1}{2} \int_{-1}^{1} \gamma_{\alpha\beta} \, d\zeta_3,$$
$$\tau_\alpha = \frac{1}{2} \int_{-1}^{1} \chi_{3\alpha} \, d\zeta$$

The components of Cosserat plate strain can also be represented in terms of the components of set $\mathcal{U}$ by the following strain-displacement relation of the Cosserat plate

$$v_{\alpha\beta} = \hat{\Psi}_{\beta,\alpha} + \varepsilon_{3\alpha\beta} \hat{\Omega}_3,$$
$$\tau_\alpha = \hat{\Omega}_{3,\alpha}$$

Now, we are able to postulate the HPR variational principle for the twisting of Cosserat plate P.

Let $\mathcal{A}$ denote the set of all admissible states that satisfy the Cosserat plate strain-displacement relations and let Θ be the functional on $\mathcal{A}$ defined as

$$\Theta(s) = (U_K + K_T)$$
$$- \int_{P_0} (\mathcal{S}_T \cdot \mathcal{E}_T + v_1 \hat{\Omega}_3) da$$
$$+ \int_{\Gamma_\sigma} \mathcal{S}_T^o \cdot \left(\mathcal{U}_T - \mathcal{U}_T^o\right) ds$$
$$+ \int_{\Gamma_u} \mathcal{S}_T^n \cdot \mathcal{U}_T ds,$$

for every $s = [\mathcal{U}_T, \mathcal{E}_T, \mathcal{S}_T] \in \mathcal{A}$.
  Then

$$\delta\Theta(s) = 0$$

is equivalent to the following plate twisting equilibrium system of equations

$$N_{\alpha\beta,\alpha} = \hat{\rho}\frac{\partial^2 \hat{\Psi}_\alpha}{\partial t^2}, \qquad (B.27)$$

$$M^*_{\alpha,\alpha} + \epsilon_{3\beta\gamma} N_{\beta\gamma} + v_1 = \hat{I}_3 \frac{\partial^2 \hat{\Omega}_3}{\partial t^2}, \qquad (B.28)$$

where

$$\hat{\rho} = \rho h,$$
$$\hat{I}_3 = J_{33} h,$$

with the resultant traction boundary conditions at Γ_σ:

$$N_{\alpha\beta} n_\beta = N_{o\alpha},$$
$$M^*_\alpha n_\alpha = M^*_o,$$

and the resultant displacement boundary conditions

$$\mathcal{U}_T^o = \left[\hat{\Psi}_{o\alpha}, \hat{\Omega}_{o3}\right]$$

at Γ_u

$$\hat{\Psi}_\alpha = \hat{\Psi}_{o\alpha},$$
$$\hat{\Omega}_3 = \hat{\Omega}_{o3}.$$

and the constitutive formulas

$$v_{\alpha\alpha} = \frac{\partial \Phi}{\partial N_{\alpha\alpha}} = \frac{\lambda + \mu}{h\mu(3\lambda + 2\mu)} N_{\alpha\alpha} - \frac{\lambda}{2h\mu(3\lambda + 2\mu)} N_{\beta\beta},$$

$$v_{\alpha\beta} = \frac{\partial \Phi}{\partial N_{\alpha\beta}} = \frac{\alpha + \mu}{4h\alpha\mu} N_{\alpha\beta} + \frac{\alpha - \mu}{4h\alpha\mu} N_{\beta\alpha},$$

$$\tau_{\alpha} = \frac{\partial \Phi}{\partial M_{\alpha}^*} = \frac{\gamma + \epsilon}{4h\gamma\epsilon} M_{\alpha}^*.$$

We also assume that the initial condition can be presented in the form

$$\mathcal{U}_T(x_1, x_2, 0) = \mathcal{U}_T^0(x_1, x_2),$$

$$\frac{\partial \mathcal{U}_T}{\partial t}(x_1, x_2, 0) = \mathcal{V}_T^0(x_1, x_2)$$

Appendix C

MINDLIN-TYPE CONSTITUTIVE FORMULAS

In order to be consistent with the Mindlin-Reissner Plate Theory, the constitutive formulas are modified as follows

$$M_{\alpha\alpha} = \frac{h^3\mu(\lambda+\mu)}{3(\lambda+2\mu)}\Psi_{\alpha,\alpha} + \frac{\lambda\mu h^3}{6(\lambda+2\mu)}\Psi_{\beta,\beta} + \frac{\left(6p^*+\pi^2\hat{p}\right)\lambda h^2}{60(\lambda+2\mu)},$$

$$M_{\beta\alpha} = \frac{(\mu-\alpha)h^3}{12}\Psi_{\alpha,\beta} + \frac{h^3(\alpha+\mu)}{12}\Psi_{\beta,\alpha} + (-1)^\beta\frac{\alpha h^3}{6}\Omega_3,$$

$$Q_\alpha = \frac{\pi^2 h(\alpha+\mu)}{12}\Psi_\alpha + \frac{\pi^2(\mu-\alpha)h}{12}W_{,\alpha} + \frac{2(\mu-\alpha)h}{3}W^*_{,\alpha}$$
$$+ (-1)^\beta\frac{\pi^2 h\alpha}{6}\Omega_\beta + (-1)^\beta\frac{h\pi^2\alpha}{6}\hat{\Omega}_\beta,$$

$$Q^*_\alpha = \frac{\pi^2(\mu-\alpha)h}{12}\Psi_\alpha + \frac{\pi^2(\mu-\alpha)^2 h}{12(\mu+\alpha)}W_{,\alpha} + \frac{2(\mu+\alpha)h}{3}W^*_{,\alpha}$$
$$+ (-1)^\alpha\frac{\pi^2 h\alpha}{6}\Omega_\beta + (-1)^\alpha\frac{\pi^2 h\alpha(\mu-\alpha)}{6(\mu+\alpha)}\hat{\Omega}_\beta,$$

$$\hat{Q}_\alpha = \frac{8\alpha\mu h}{3(\mu+\alpha)}W_{,\alpha} + (-1)^\alpha\frac{8\alpha\mu h}{3(\mu+\alpha)}\hat{\Omega}_\beta,$$

$$R^*_{\beta\alpha} = \frac{\pi^2(\gamma-\epsilon)h}{12}\Omega_{\beta,\alpha} + \frac{\pi^2 h(\gamma+\epsilon)}{12}\Omega_{\alpha,\beta},$$

$$R^*_{\alpha\alpha} = \frac{\pi^2 h\gamma(\beta+\gamma)}{3(\beta+2\gamma)}\Omega_{\alpha,\alpha} + \frac{\pi^2 h\beta\gamma}{6(\beta+2\gamma)}\Omega_{\beta,\beta},$$

$$\hat{R}_{\beta\alpha} = \frac{2(\gamma-\epsilon)h}{3}\hat{\Omega}_{\beta,\alpha} + \frac{2(\gamma+\epsilon)h}{3}\hat{\Omega}_{\alpha,\beta},$$

$$\hat{R}_{\alpha\alpha} = \frac{8\gamma(\gamma+\beta)h}{3(\beta+2\gamma)}\hat{\Omega}_{\alpha,\alpha} + \frac{4\gamma\beta h}{3(\beta+2\gamma)}\hat{\Omega}_{\beta,\beta},$$

$$S_\alpha = \frac{\pi^2\gamma\epsilon h^3}{6(\gamma+\epsilon)}\Omega_{3,\alpha}.$$

Bibliography

[1] Thermal insulation materials made of rigid polyurethane foam (pur/pir). *BING Report*, 1, 2006.

[2] Green A. and Naghdi P. The linear theory of an elastic cosserat plate. *Mathematical Proceedings of the Cambridge Philosophical Society*, 63:537–550, 1966.

[3] Khanna A. and A. Sharma. Mechanical vibration of visco-elastic plate with thickness varation. *International Journal of Applied Mathematical Research*, 1:150–158, 2012.

[4] Merkel A., Tournat V., and Gusev V. Experimental evidence of rotational elastic waves in granular phononic crystals. *Phys. Rev. Lett*, 107: 225502, 2011.

[5] Rossle A., Bischoff M., Wendland W., and Ramm E. On the mathematical foundation of the (1,1,2)-plate model. *International Journal of Solids and Structures*, 36:2143–2168, 1999.

[6] Botelho E. C., Silva R. A., Pardini L. C., and Rezende M. C. A review on the development and properties of continuous fiber/epoxy/aluminum hybrid composites for aircraft structures. *Materials Research*, 9:247–256, 2006.

[7] Castelain C., Mokrani A., Legentilhomme P., and Peerhossaini H. Residence time distribution in twisted pipe flows: helically coiled system and chaotic system. *Experiments in Fluids*, 22:359–368, 1997.

[8] Eringen A. C. Theory of micropolar plates. *Journal of Applied Mathematics and Physics*, 18:12–31, 1967.

[9] Johnson C. *Numerical Solution of Partial Differential Equations by Finite Element Method.* Cambridge University Press, 1987.

[10] Chladni E. *Entdeckungen uber die Theorie des Klanges.* 1787.

[11] Reissner E. On the theory of elastic plates. *Journal of Mathematics and Physics*, 23:184–191, 1944.

[12] Reissner E. The effect of transverse shear deformation on the bending of elastic plates. *Journal of Applied Mechanics*, 12:69–77, 1945.

[13] Ventsel E. and Krauthammer T. *Thin Plates and Shells*. Marcel Dekker, Inc., first edition, 2001.

[14] Kima H., Junga D., Jungb I., Cifuentesb J., Rheeb K., and Huib D. Enhancement of mechanical properties of aluminium/epoxy composites with silane functionalization of aluminium powder. *Composites: Part B*, 43, 2012.

[15] Love A. E. H. On the small free vibrations and deformations of elastic shells. *Philosophical trans. of the Royal Society (London)*, A(17):491–549, 1888.

[16] Love A. E. H. *A Treatise on the Mathematical Theory of Elasticity*. Cambridge University Press, 1892.

[17] Adrianov I. and Awrejcewicz J. Theory of plates and shells: New trends and applications. *International Journal of Nonlinear Sciences and Numerical Simulations* 5(1):23–36, 2004.

[18] Bernoulli J. Essai théorique sur les vibrations des plaques élastiques rectangularies et libres. *Nova Acta Acad Petropolit*, 5:197–219, 1789.

[19] Reddy J. *Theory and Analysis of Elastic Plates and Shells*. CRC Press, Taylor and Francis, 2007.

[20] Conway J.B. *A Course in Functional Analysis*. 1985.

[21] Altmann S. L. *Rotations, Quaternions and Double Groups*. Oxford, University Press, 1986.

[22] Donnell L. *Beams, Plates and Shells (Engineering societies monographs)*. McGraw-Hill Inc, 1976.

[23] Euler L. De motu vibratorio tympanorum. *Novi Commentari Acad Petropolit*, 10:243–260, 1766.

[24] Evans L. *Partial Differential Equations*. 2010.

[25] Steinberg L. Mesoelastic deformation with strain singularities. *Mathematics and Mechanics of Solids*, 11(4):385–400, 2006.

[26] Steinberg L. Deformation of micropolar plates of moderate thickness. *International Journal Applied Mathematics and Mechanics*, 6(17):1–24, 2010.

[27] Steinberg L. and Kvasov R. Enhanced mathematical model for cosserat plate bending. *Thin-Walled Structures*, 63:51–62, 2013.

[28] Steinberg L. and Kvasov R. Analytical modeling of vibration of micropolar plates. *Applied Mathematics*, 6:817–836, 2015.

[29] Steinberg L. and Kvasov R. Distinctive characteristics of cosserat plate free vibrations. *Dynamical Systems Theory*, 6:817–836, 2019.

[30] Gurtin M. *The Linear Theory of Elasticity*. Springer-Verlag, 1972.

[31] Morley M. *Building with Structural Insulated Panels*. Taunton Press, 2000.

[32] Surappa M. Aluminium matrix composites: Challenges and opportunities. *Sadhana*, 28:319–334, 2003.

[33] Madrid P. *Reissner Plate Theory in the Framework of Asymmetric Elasticity*. University of Puerto Rico, 2007.

[34] Neff P. Relations of constants for isotropic linear Cosserat elasticity. *Zeitschrift fur Angewandte Mathematik und Mechanik*, 2008.

[35] Neff P. and Jeong J. A new paradigm: the linear isotropic Cosserat model with conformally invariant curvature energy. *Zeitschrift fur Angewandte Mathematik und Mechanik*, 89:107–122, 2009.

[36] Gauthier R. Experimental investigations on micropolar media. *Mechanics of Micropolar Media*, 1:395–463, 1982.

[37] Gauthier R. and Jahsman W. A quest for micropolar elastic constants. *Journal of Applied Mechanics*, 42:369–374, 1975.

[38] Kumar R. Wave propagation in micropolar viscoelastic generalized thermoelastic solid. *International Journal of Engineering Science*, 38, 2000.

[39] Kumar R. and Gupta R. Propagation of waves in transversely isotropic micropolar generalized thermoelastic half space. *International Communications in Heat and Mass Transfer*, 37:1452–1458, 2010.

[40] Kvasov R. Mathematical modeling and finite element computation of cosserat elastic plates. *Ph.D. Thesis*, 2013.

[41] Kvasov R. and Steinberg L. Numerical modeling of bending of cosserat elastic plates. *Proceedings of the 5th Computing Alliance of Hispanic–Serving Institutions Annual Meeting*, San Juan, Puerto Rico, pages 67–70, 2011.

[42] Kvasov R. and Steinberg L. Numerical modeling of bending of micropolar plates. *Thin-Walled Structures*, (69):67–78, 2013.

[43] Kvasov R. and Steinberg L. Modeling of size effects in bending of perforated cosserat plates. *Modelling and Simulation in Engineering*, 2017(1–19), 2017.

[44] Lakes R. Experimental microelasticity of two porous solids. *International Journal of Solid Structures*, 22(1):55–63, 1986.

[45] Lakes R. Experimental methods for study of Cosserat elastic solids and other generalized elastic continua. *Continuum models for materials with micro-structure*, 1:1–22, 1995.

[46] Mindlin R. Micro-structure in linear elasticity. *Archive of Rational Mechanics and Analysis*, 16:51–78, 1964.

[47] Nagle R., Saff E., and Snider A. *Fundamentals of Differential Equations and Boundary Value Problems*. 5th edition, Pearson, 2008.

[48] Purasinghe R., Tan S., and Rencis J. "Micropolar elasticity model for stress analysis of human bones," Images of the Twenty-First Century. *Proceedings of the Annual International Engineering in Medicine and Biology Society*, 1989, pp. 839–840 vol.3, doi: 10.1109/IEMBS.1989.96010.

[49] Reyes R. Comparison of elastic plate theories for micropolar materials. *M.S. Thesis*, 2010.

[50] Singh S. Blowing agents for polyurethane foams. *Rapra Review Report*, 12, 2002.

[51] Timoshenko S. and Woinowsky-Krieger S. *Theory of Plates and Shells*. McGraw-Hill, 1959.

[52] Chang T., Lin H., Wen-Tse C., and Hsiao J. Engineering properties of lightweight aggregate concrete assessed by stress wave propagation methods. *Cement and Concrete Composites*, 28:57–68, 2006.

[53] Hughes T. *The Finite Element Method: Linear Static and Dynamic Finite Element Analysis*. Prentice-Hall, 1987.

[54] Gerstle W., Sau N., and Aguilera E. *Micropolar Peridynamic Constitutive Model for Concrete*. 2007.

[55] Nowacki W. *Teoria Sprezystosci*. Panswowe Wydawnictwo Naukowe, 1970.

[56] Pabst W. Micropolar materials. *Ceramics*, 49:170–180, 2005.

Index